TRACKING

A Blueprint for Learning How

by

Jack Kearney

Pathways Press El Cajon

Cover Photo by
Joe O'Dell

Fourth Printing, April 1986

Pathways Press El Cajon

Library of Congress No: 77-18472

Manufactured in the United States

DEDICATION

This book is lovingly dedicated to my wife, Avis, without whose untiring efforts it could never have been accomplished. Her assistance covered grammar, sentence structure, punctuation, and typing,...tons of typing. Throughout the months of writing, re-writing, typing, re-typing, correcting, changing, and typing again, she maintained her same even disposition: whining, complaining, and footdragging every tortuous inch of the way.

ACKNOWLEDGMENTS

Many people contributed in some way to the eventuality of this book. However, the three people who invested the greatest amount of time and effort are my fellow San Diego Mountain Rescue Team members: John Wehbring, Bill Mackintosh, and Joe O'Dell. John took my original manuscript and promptly demolished it. As the pieces were put back together it began to resemble a book. Had I followed all of his suggestions the final product might even have looked professional. Bill Mackintosh not only proofread and offered suggestions on the text but took photographs and handled all film developing, enlarging, printing, and cropping. Joe O'Dell spent innumerable hours, took long trips, and probably took a half dozen days from his job in order to set up and shoot precisely the picture I wanted in order to illustrate a particular point. With these three people I particularly wish to share the credit,...or the blame.

In the last ten years many good trackers have worked at the El Cajon Border Patrol Station and all of them helped to build our envious record of success on searches for lost persons. However, in some outlandish places and at some outrageous hours, the guy I looked up and saw the most often was Jim Burns, and only slightly less often was Larry Harlan.

The initial motivating force behind our involvement in Search and Rescue work was our boss, Ab Taylor. He, more than anyone else, has been beating the drum and selling the idea of tracking as a rescue tool. It was originally his idea to try teaching tracking to search volunteers and his dedication to this pursuit continues to occupy a major portion of his time.

Others who provided vital input, encouragement, or inspiration were Lois McCoy, Jon Gunson, PeeWee Lagasse, Stan Bush, and the San Diego Mountain Rescue Team.

Jack Kearney

TABLE OF CONTENTS

1

Man, The Tracker

Approximately one million years ago there walked upon the earth, for the first time, a creature called "Homo Erectus", a creature that scientists are willing to agree was man.

Another million years prior to Homo Erectus there existed a man-like creature known as "Austrolopithecus Africanus" who is believed to be the direct forebear of Homo Erectus. Austrolopithecus is special because he is the earliest of our ancestors to leave us evidence of his tools, the crude but sharp-edged stone tools of a meat eater.

The fact that Austrolopithecus was a meat eater tells our archaeologists that he was a hunter and thus distinguishes him from all of his predecessors who were believed to be browsers and grazers.

The fact that Austrolopithecus hunted means he had to seek out and stalk his food source. We can only guess how he did this, but since many of the lower forms of animal hunt by scenting, this pre-man probably did also. Surely, very early in this period of Man's development as a hunter, he used his vision to assist him. He must have looked for evidence of game and become aware of sign left on the ground by the movement of the animals he sought. He soon probably associated these tracks with food and began to look for them when hunting. Thus his skill in tracking probably evolved at about the same time as did his earliest use of tools. It is therefore reasonable to assume that the development of tracking as a method of locating something you are seeking probably goes back so far in pre-history as to antedate the presence of Man himself.

As man became more civilized and less dependent on hunting as a method of securing food, his skill in tracking undoubtedly deteriorated. However, it never completely vanished and one need only look into American history to find tracking a flourishing art as recently as one hundred years ago.

Much of our current awareness of tracking derives from Hollywood's romantic versions of the winning of the West. These exciting western movies showed us how the daring settlers knew when the Indians were near by keeping track of footprints on the trails and around the watering holes. We learned that both the Indian and early settler relied on evidence of tracks to tell them if game was available and how plentiful it was. We learned that a number of Indian tribes regularly moved their encampments to follow their food source. However, the most vivid tracking lessons we learned were from the cowboy movie's most heroic character, the scout of the wagon train. This eagle-eyed hero could take one look at the ground and know everything that had happened between the Mississippi River and the Pacific Ocean for the past eight days, why it had happened, and who was going to get the blame.

TRACKING IN THE BORDER PATROL

It generally comes as quite a surprise when people discover that there are still men who practice this ancient art with as much skill and frequency as did the early cowboy and Indian. These men not only track other human beings almost daily, but actually earn their living at it. The United States Border Patrol has been charged with the responsibility of apprehending aliens who try to enter the United States illegally, and the practice of tracking these aliens, of actually following their "sign" through the rugged back country, is a technique that is as old as the Border Patrol itself.

In 1911 when federal officers first began to patrol the Mexican Border the original seventy-five man force was made up largely of ex-cowboys. Almost all of these men were from the Southwestern United States where ranching and the daily tracking of cattle were a way of life. These early officers were not so far removed in time or spirit from this country's period of great western migration that they had forgotten that tracking was a skill needed for survival by everyone in the early days of the West. Therefore, it is easily understandable that these men would resort to tracking in order to apprehend aliens who tried to walk into the United States undetected. Thus tracking became the earliest Border Patrol technique and the continued emphasis on its use has caused the Border Patrol to develop the largest pool of expert trackers that exists in the United States today.

It is wrong to assume, however, that all Border Patrol Agents are expert trackers. Probably 90 per cent of all Border Patrol Agents are less than expert due to the fact that they are not involved in an activity that allows them to perfect this unusual skill. The largest concentrations of our officers are assigned to "linewatch" activities and are stationed immediately along the international boundary opposite large Mexican and Canadian border cities.

These cities are major transportation hubs and provide

a jumping-off place for most aliens who try to enter the country illegally. Therefore, most agents spend their time "on the line" opposite such Mexican cities as Tijuana, Mexicali, Nogales, Juarez, and Matamoros or across from such Canadian cities as Vancouver, Winnipeg, Toronto, Montreal, and Quebec. Officers at these line stations are primarily concerned with observing and apprehending aliens in the act of entering, and rely on "back-up" stations to pursue those aliens who have eluded this first line of defense.

Still, other Border Patrol officers are assigned to stations that are many miles from the border but are situated near major highways or railways so as to facilitate the apprehending of aliens on their way to the interior. These stations are located in such places as San Clemente, California; Temecula, California; Indio, California; Lourdsburg, New Mexico; and Miami, Oklahoma. Officers at these stations are engaged

Border Patrol agents checking traffic at San Clemente, California.

in train check or traffic check and have little opportunity to develop expertise in tracking.

Quite a number of Border Patrol Stations are located in areas that attract alien workers who are in the country illegally. Since most of these stations were established during that period of time when the largest number of illegal alien workers sought work in agriculture, these stations are situated near areas of high agricultural activity. These stations can be found in such places as Sacramento, California; Twin Falls, Idaho; Amarillo, Texas; Little Rock, Arkansas; and Lake Charles, Louisiana. These stations are so many miles from a land border that it is quite obvious the officers at these stations do not spend their time tracking people who are trying to sneak into the country.

It is predominantly at back-up stations like El Cajon, California, or at small line stations where few illegal entries occur that Border Patrol Officers have the opportunity to become highly skilled trackers.

TRACKING AT THE EL CAJON BORDER PATROL STATION

In 1948 the Border Patrol established a station at El Cajon, California with a nucleus of four good trackers. Located just east of San Diego, the purpose of the station was to back up the line stations of Campo and Chula Vista. In the 1950's there were changes in personnel but the number of officers changed very little. With my arrival in 1961 the station was increased to six men and in the intervening years has grown to a current complement of about 20 officers. It is still a relatively small station but its efficiency (the ratio of aliens apprehended to man hours expended) is perennially among the highest in the United States. It is a rare month that fewer than 1,200 aliens are apprehended and our peak months average close to 3,000 persons per month. Nearly 60 per cent of these are caught as a result of tracking.

The El Cajon Border Patrol Station's area of responsibility is uniquely shaped and situated so as to almost force its complement of officers to become expert trackers in order to perform their duties. Located in San Diego County, where in 1976, 25 per cent of all the undocumented aliens in the United States were apprehended, it is an area of high activity. The station is shaped somewhat like a spreading tree with its narrow trunk setting on the Mexican Border between Otay Peak and Tecate Peak. It grows upward and outward into a secondary line of defense to the north of the Campo and Chula Vista stations. Functioning much like a linebacker in football who moves to pick up the rushes that elude his linemen, the El Cajon Station is responsible for taking the second crack at those people who have eluded detection and managed to get several miles into the United States. With over 1,000 square miles of area to patrol, much of it rugged and mountainous, the El Cajon tracker may be able to stalk his quarry for upwards of fifty miles. Over the years I have worked tracks on a number of chases that lasted as long as three days and fondly remember one that traversed almost one hundred miles!

My own first exposure to tracking began when I entered the Border Patrol in 1960. However, since I was assigned to a line station I really did not start acquiring skill as a tracker until I transferred to El Cajon in 1961. At that time, I began tracking from four to ten hours a day, three to five days a week and have continued to do so for a period of time that is now approaching twenty years. This is, no doubt, a solid background of experience for a tracker, but sometimes the quality of the experience can be even more valuable than the quantity. The interesting "catch" to a Border Patrolman's experience is that every blessed person he goes after, HIDES. Getting within shouting distance of my quarry when he is awake, alert, and listening gets me an absolute zero—if he can hear me coming he will usually cut and run before I get there. I may have to track him step-by-step

Lightened section indicates area of responsibility of U.S. Border Patrol station, El Cajon, California.

to the exact bush under which he is hiding, crawl under the bush, and dig him out from under the leaves with which he has concealed himself. "Close" for me is failure.

There is a high element of sport in tracking and I do not have to spend much time in reflection to realize that over the years my tracking skill has brought me more adventure, excitement, satisfaction, and downright fun than anyone should expect from something they get paid to do. Being a Border Patrol Agent is not exactly an everyday, mundane type of job. Dealing with smugglers and felons can be dangerous and risky and tends to keep a person alert and interested. However, if you have the added dimension of being a tracker, you will find yourself being called upon to assist other law enforcement agencies in searches for burglers, rapists, murderers, kidnappers, escaped prisoners, and arsonists. Yet, the most satisfying tracking experience you are apt to encounter is the chance to search for, and find, a person who is lost in the wilderness.

TRACKING THE LOST PERSON

It is difficult to pinpoint exactly when I began to get involved in search and rescue work. I can remember teaching tracking to local volunteers in the late 1960's, but I don't remember actually participating in a search until about 1970. The reputation built by El Cajon officers as skillful trackers of aliens caused us to receive requests for instruction in our tracking techniques. As search and rescue personnel were exposed to this instruction they began to realize the value of an expert tracker and we were requested to assist on actual searches. It was from these vague beginnings that we embarked on our enviable record of live "finds" of lost children which has propelled us to national prominence.

Prior to 1967 I was unfamiliar with searches for lost persons but, knowing the value of tracking, I had supposed that all searchers were similarly informed and therefore accomplished trackers. It came as quite a shock when I learned that with most search and rescue groups tracking was a rudimentary tool at best. Upon reflection, it was understandable that the average person would seldom have the time, opportunity, or motivation to learn this painstaking skill. What was much slower in coming was the realization that people are almost completely unaware of what an effective method tracking is for finding lost persons and, in fact, have no idea how often it is the BEST method.

The San Diego Mountain Rescue Team was one of the first groups to receive our training and is chiefly responsible for spreading the word of our successful "step-by-step" teaching technique. Our two-day tracking indoctrination course is sought after, and has been taught to SAR and law enforcement groups from all over the country, and has been presented to groups from as far away as Guam. It is acclaimed because it works. We teach it because the most common problem we trackers are forced to cope with is the confusing tracks laid down by non-trackers in a search area.

The author conducting field training for search and rescue group.

I have been on a number of searches that found non-trackers clomping randomly through an area because they decided the area would not reveal tracks and therefore, they felt no harm could be done. I could have easily followed tracks in the area and so could at least eight or nine of my fellow officers at El Cajon. I have heard or read of countless other searches where the subject was never found, or found dead, in terrain where our tenth or eleventh best tracker could have easily found him alive in eight hours, if he could have gone in alone before the area was trampled by well-meaning but short-sighted amateurs.

To a non-tracker this may sound like outrageous boasting, but it is based on years of experience and sound confidence in our men. Daily, routinely, for long distances we follow aliens who illegally cross the border, tracking them to their hiding places and bringing them back to the station without raising the eyebrows of fellow officers. It is our job. Every man at El Cajon does it, with varying degrees of skill. Border Patrolmen at other stations do it and have been doing it every working day for generations. Outsiders may marvel, but within the Border Patrol it is not considered noteworthy enough to elicit comment.

None of us are clairvoyant; none of us are blessed with extra sensory perception. We are not unusual people with eyes of eagles, hearts of lions, and souls of saints. Not one of us was raised by the Shoshone or suckled by timber wolves. We are an unimposing gaggle of bandy-legged, pot-bellied liars and libertines who have become expert at what we do, by doing it daily and making a lot of mistakes.

The purpose of this book is to short-cut the learning process by eliminating those time consuming mistakes. Anyone reading this book who has 20-20 vision (with, or without glasses), patience, and enough motivation to *practice* the system outlined herein, can become just as expert as any one of us.

2

Why Tracking?

Tracking can be a great aid to any investigation because it is an outstanding information gathering process. Tracks are clues, and they are by far the most plentiful type of clue that a lost person or law violator will leave behind. To the eye of an expert tracker, a bit of evidence will appear just about every place that the person he is seeking has moved. In ideal circumstances this means a clue just about every 18 to 20 inches.

A common training exercise for beginning law enforcement officers and beginning searchers has the instructor dropping bits of evidence along a trail and then grading the students on the number of objects that are found. This trail of litter may include such things as a pencil, a gum wrapper, a cigarette butt, a coin, or a matchbook cover. It is a fun exercise and everyone usually has a very good time. However, the instructional value is comparable to teaching a person to accurately shoot a basketball by having that person begin by throwing it into a swimming pool. Over the last eighteen years I have followed sets of tracks leading to the finding of tens of thousands of people and certainly I have found things that they have discarded or lost. But, on the average, the finding of this kind of litter only occurs about once in every three miles of tracking.

On the other hand, the information that can be assembled by a tracker who finds no other evidence than tracks can sometimes recreate an entire chain of events.

Several years ago El Cajon Border Patrol Officers assisted on a murder case in which an attractive young lady was found stabbed to death within fifty feet of her automobile which had broken down on a deserted dirt road. The circumstances posed many questions: Had she encountered car trouble and while walking to get assistance had she been accosted by a stranger who happened to be lurking in the area? Had she been kidnapped and brought to the area to be killed? Had she come to the area with someone she knew? All of these possibilities and many others existed initially. Fortunately the agency having jurisdiction called our El Cajon office immediately and requested tracking assistance. We responded promptly and within two hours we could tell the following things with a fair degree of certainty:

> The female had been driving at the time the car became disabled. The murderer had been riding as her passenger. When the car stopped, the female had gotten out and walked towards the front of her vehicle while her passenger, whose tracks indicated that he was an adult male, had departed from the passenger side and walked around the rear of the car giving the victim an excellent chance to run had she felt threatened. The male and female milled around the car a bit then started walking a dirt road towards the top of a hill. During this walk of nearly three hundred yards, the

two people walked on opposite sides of the road, meaning that there was at least a six foot separation between them even if they had been walking abreast. However, this was not always the case; at the steepest part of the hill the female lagged behind and crossed over to walk in the footsteps of the male. Again this was strong evidence that she was not being coerced. Once atop the hill they viewed the surroundings, made some kind of decision, and returned in the direction of the car. On this return trip they stopped twice prior to reaching the car and had some kind of face-to-face encounter, a struggle ensued, and then the killing. The murderer then departed the area to the west where his tracks were lost on the grass lawns and paved streets of a housing area.

The information that the trackers supplied did not solve this case, but it made a great contribution to the eventu-

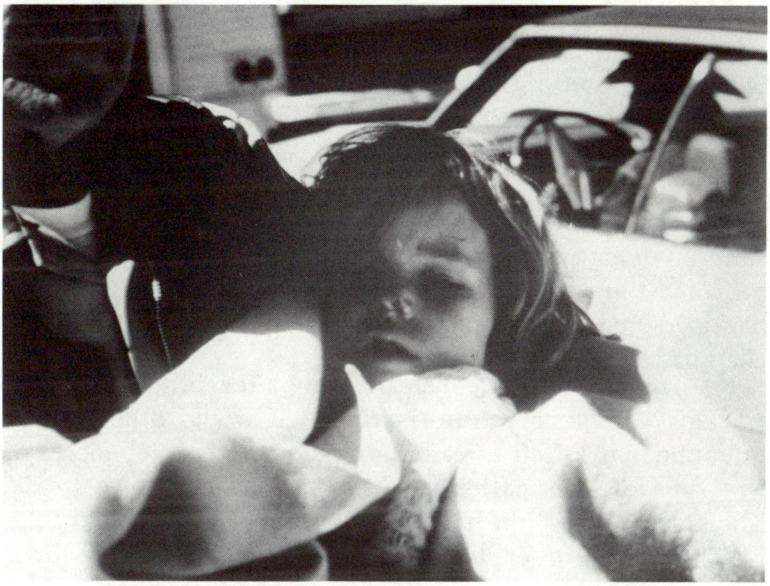

Why tracking? Because of this cherished memory from Sequoia National Forest.

al solving of it. When the young man was brought to trial this information supported his contention that the crime was not premeditated but had resulted from his losing his temper at taunts that had been thrown at him during an empassioned argument.

On other occasions I have gotten a great deal of insight into the way a person was thinking by information that I have gathered as a result of following his tracks. The fact that the tracks show the person being followed made a sudden turn and took a circuitous route around a parked vehicle shows he had not wanted to be seen by the vehicle's occupants. The fact that the subject departs from a trail and hides, indicates that he probably did not want to be seen by others walking along the trail.

At times I have had the track I was following meet with the tracks of another person, the number and position of the sets indicating that a conversation probably took place. On some of these occasions the second set of tracks have returned to a nearby farm house, thereby providing me with the opportunity to talk to someone who has seen and talked to the person I had been following.

Sometimes, in following the tracks of aliens who have entered the United States illegally, I have passed close enough to a farm house so as to arouse the barking of dogs. By simply talking with the occupants of the house to determine at what time the dogs had been provoked into a similar barking spree I have been able to pinpoint the time that the alien passed by.

These examples constitute only a few of the ways that tracking can uncover information that would otherwise be unobtainable. Often the addition of only a few key pieces of this type of information has been sufficient to solve a case. The tracker does not have to follow a set of tracks to their conclusion in order to prove his value; often if he does nothing more than positively identify which track belongs to the person being sought he has made a tremendous contribution.

I can remember quite a number of searches that accomplished nothing for several days and, once the victim's track was identified, plummeted to a successful conclusion within hours. Once the correct track description was broadcast over the radio other searchers remembered seeing the print, and a travel pattern quickly emerged that enabled a trail to be plotted to the victim's location.

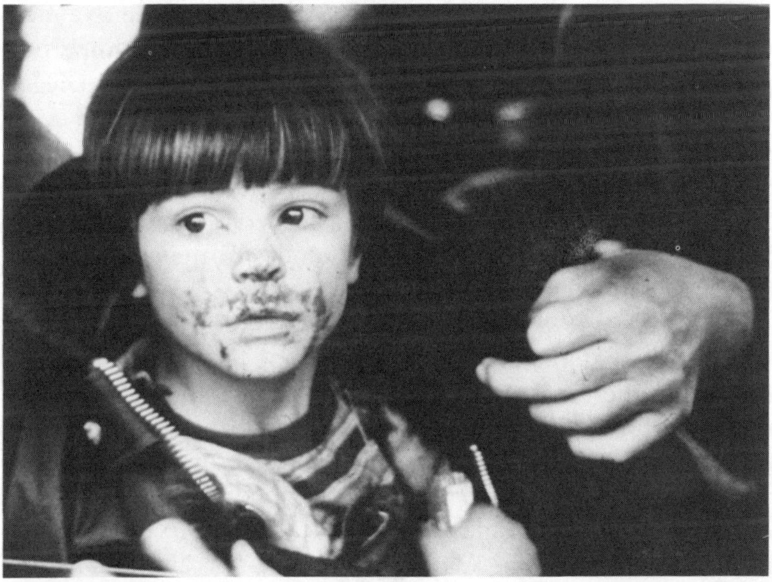

Why tracking? Because of this cherished memory from Indio, California.

In many cases tracking will allow you to focus your search in a rather confined area. At other times it can eliminate areas that do not need to be searched and often it is a tool by which you may disprove, or verify, information that you have uncovered through interrogation of witnesses.

An example of this occurred a number of years ago when I arrested the driver of a vehicle who was transporting five illegal aliens into the United States. Interrogation of these five Mexican men revealed that they had made contact with this man in Mexico, had made arrangements with him to be smuggled to Los Angeles for $200 apiece, had entered his car in Mexico (two of them having been put in the trunk), and had been driven into the United States in a remote area thru a hole the driver had cut in the International fence. Subsequent interrogation of the driver elicited his completely different version of the story; he had made a rest stop at a small state campground about forty miles east of San Diego (and many miles inside the United States) and these five men had come up out of a dry river bed area, leaving numerous companions behind, and asked him for a ride to San Diego.

The smuggler told his story so convincingly and with such attention to subtle detail that I began to wonder if there really might be more aliens in the area where he had stopped. My curiostity was so aroused that after wrapping up the case, I got in my car and drove the forty miles to the campground and thoroughly checked the entire area.

I discovered not only that there were no additional aliens, but that the only tracks that came up out of that river bed were the tracks of two people, a male and a female. It could be further ascertained that the tracks had originated in the campground in the first place and were simply returning, and the condition of the tracks indicated that this had occurred more than two days earlier. In addition to this I found no tracks of the five Mexican citizens, nor any tracks belonging to the smuggler anywhere in the area.

Despite the fact that we had a very strong case, our position was strengthened at the trial when I was able to testify with conviction to the facts outlined above. The smuggler's defense was greatly discredited by this testimony and he was convicted on all counts.

Probably the strongest case that can be made for the use of tracking is the one we regularly present in our training lectures and which is most easily explained in simple mathematical terms: A soldier marching in formation, at attention, travels at a rate of four miles an hour. In military parlance this pace is called "quick time" and is admittedly quite a bit faster than a leisurely stroll. One-half this speed, or two miles an hour, is certainly within the realm of possibility of all but the very young or the very old. Difficult terrain could reduce walking speed even further, but, for planning purposes, two miles an hour is reasonable.

Assuming that a lost person is walking at a speed of two miles an hour, it is theoretically possible, if he has been missing for five hours, for him to have traveled ten miles from the point where he was last seen. Since he could have traveled in any direction, you are contending with a circular area that has a 20 mile diameter and encompasses 314 square miles.

Since full scale searches are seldom under way until a person has been missing for at least ten hours, there exists a potential search perimeter having a 20 mile radius, a 40 mile diameter, and an area of 1,256 square miles before the first trained searcher arrives on the scene.

As each hour passes the circle expands and in the tenth hour of a full-scale search (twentieth hour since the victim walked away) you are presented with a search area of 5,026 square miles.

The number of people required to thoroughly search an area of this size is staggering. In a heavily forested area a search group numbering one hundred people could take several hours to sweep search a single square mile of this type of difficult terrain.

If a tracker does nothing more than establish the first five steps taken by the lost person he will have established the probability of a direction that immediately cuts the search area in half. If he can maintain the direction of the lost person's trail for as much as a mile, the shape of the search area can be reduced from a 360 degree circle to a

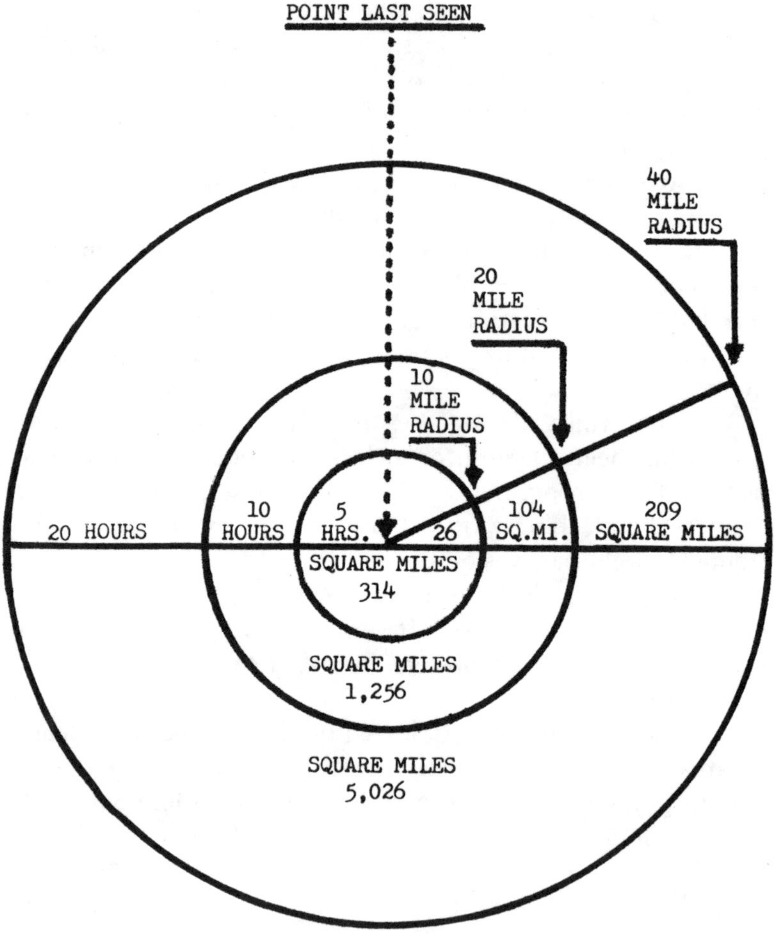
POINT LAST SEEN
40
MILE
RADIUS
20
MILE
RADIUS
10
MILE
RADIUS
20 HOURS
10
HOURS
5
HRS.
26
104
SQ.MI.
209
SQUARE MILES
SQUARE MILES
314
SQUARE MILES
1,256
SQUARE MILES
5,026

pie-shaped wedge that is likely to be no greater than 30 degrees. This narrow focusing of the search area reduces the number of square miles to be searched in the smaller area (5 hours of walking) from 314 square miles to 26 square miles and in the larger area (20 hours of walking) from 5,026 square miles to 209 square miles.

This, of course, is mathematical theory. How does it work out in actual search situations? It has been my experience that it holds up amazingly well.

In the preponderance of searches on which I have assisted, we have, in a short period of time, aimed the search as surely as you would aim a rifle. With the direction established, a helicopter has usually been able to fly along this corridor, spot the victim, pick him up, and return him to base camp in a relatively short period of time. What our tracking usually accomplishes is the concentrating of searchers in a small area with a high probability for containing the victim.

On these successful occasions the trackers are usually the last ones to come up out of the canyon and be transported back to base camp. Often the exhuberance of the celebration is beginning to die down and many times the trackers even miss out on the satisfaction of viewing the object of their search who may have been whisked off to undergo a medical examination. Surely the television camera will have departed and the helicopter crew will have finished with their interviews.

It is during these anti-climactic winding down periods that the trackers and ground searchers slip into quiet reflection as they become aware of the sweat and begin to feel the aches and fatigue. It is a quiet time for relishing success and usually the tracker will have ample reason for feeling a deep satisfaction at what he has accomplished.

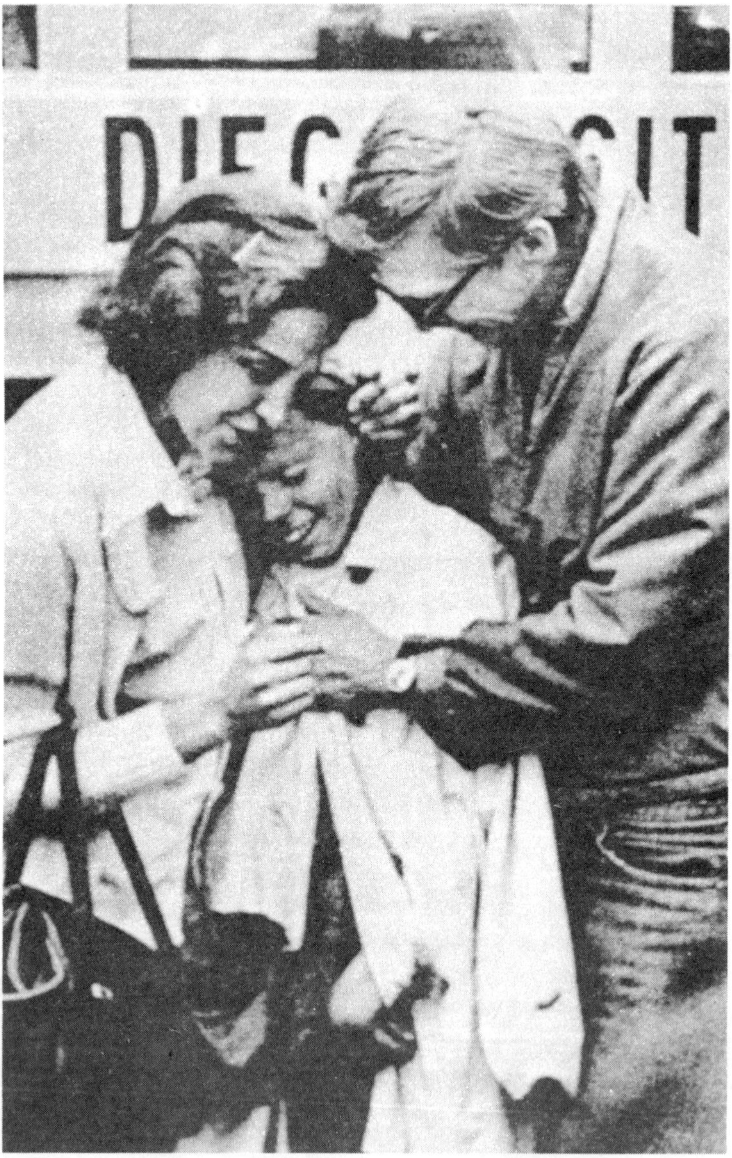

Why tracking? Because of this cherished memory from Palomar Mountain.

3

Awareness Training

Phase I, Exercise 1

It is very important that you understand the basic purpose of this book. It is to teach you how to become a tracker.

An ultimate goal, a long range purpose, is to allow you to find a lost person by applying your tracking skill. However, that is not going to happen the minute you put down this book because you will not yet have acquired the tracking skill. If you are motivated to work at it, you will become a skilled tracker; if you are not sufficiently motivated we will still probably reap the fringe benefit of reducing the likelihood of your ignorance contributing to the unnecessary death of a lost person.

Tracking is no more difficult to learn than playing the piano. However, it is not one iota easier either! Eight year old children can learn to play the piano and they can learn to track; however, both are going to require dedication to fundamentals and lots of practice.

This book is a blueprint for learning how to track. It outlines a progression of field exercises that must be followed in their proper sequence. If you follow these instructions carefully you will construct a solid foundation, follow it with a sturdy structure, and cap it off with a reliable roof. If you skip any of these vital steps you will have a very unsafe structure. If you simply read the blueprint and do not follow its instructions, you will have constructed nothing.

Preparation and Equipment

Tracking requires very little equipment beyond a pair of eyes which unaided, or aided by lenses, provide 20/20 vision. However, experience has shown that some special items can be very helpful.

Clothing depends mostly on local conditions and weather; wear whatever is normal for the area where you will be training. Sturdy clothing for brush is obvious. A broad brimmed hat is handy for shielding your eyes from sun glare, or for shading tracks at mid-day. By deliberately blocking the direct rays of mid-day sun from reaching the tracks, you create a situation that allows you to use a standard signal mirror to reflect low-angle lighting onto the shaded area. This low-angle light accentuates shadows and makes the tracks easier to see.

A walking stick is a must. It should be at least three feet long and sturdy enough to give support when negotiating steep slopes. I prefer a long metal ski pole that reaches about to my armpit. When the basket is removed from the bottom of the ski pole and two rubber bands, or "O" rings, are placed on the shaft you not only have a dependable walking stick but you will have acquired an almost magical "tracking stick."

Every tracker should carry some type of measuring device. Some trackers attach a length of metal measuring tape to their tracking stick, but most simply carry a small metal

Tracker utilizing tracking stick to determine step interval.

measuring tape in their pocket. Stiff rulers or yardsticks are undesirable because they are so unwieldy and plastic rulers break too easily in extreme cold.

A small notepad and pencil are needed for recording measurements, making notes, and making a drawing of the track you are going to follow.

Trail tape should be carried to mark tracks once you get beyond the beginning training phase (it also has other uses on searches). Plastic surveyor's flagging is popular with some search teams, but my personal preference is strips of crepe paper. The fading and deterioration of crepe paper is far faster than that of plastic and allows you to quickly differentiate old markers from fresh ones.

A last item of equipment is a flashlight. This will not be needed in the Phase I exercises but could be quite necessary if you are putting yourself through Phase II and Phase III training in heavily forested areas. Light is vital to vision.

Vision is vital to tracking. The tops of tall trees in heavily timbered areas do not allow much direct sunlight to reach the forest floor. In this dim light the use of a flashlight greatly aids tracking.

It should be mentioned that headlamps are of no value in tracking. To be effective your light should be held very low to the ground; therefore, it must be hand controlled.

Selecting a Site

The first thing you must do in learning to track or in setting up a training exercise for a group of would-be trackers, is to select a suitable training ground.

One of the early mistakes that was made in the development of our step-by-step teaching technique was to select terrain that was mildly challenging to our own abilities. We believed, incorrectly, that this would be the most exciting and interesting terrain for novices. What we did not realize was the great gap between what was obvious to a skilled tracker and what was obvious to rank beginners.

The problem is exactly like the one you face in trying to teach an adult to read. You know that they are mentally mature and cannot be stimulated by kindergarten books which have three word sentences like: "See John run"; "See Jane run". So the mistake we made in the beginning was to try to start beginners in sixth grade readers. The result was that we overwhelmed them. It was a slow trial and error process, but eventually we learned to teach to the level of capability of our students.

In the last ten years we have put on tracking seminars in very diverse locations throughout the United States and have, therefore, been forced to quickly select an appropriate training site from widely differing types of terrain.

Since our beginning exercises emphasize the importance of seeing each track in sequence we have come to be guided in our selection of a site by the thought that it must be difficult but not impossible. You should be so guided in your selection of a training site.

In the beginning exercise a set of tracks are laid out across a field and the student tracker, or trackers, attempt to follow them to their conclusion. A practice area should be selected that is flat, has a minimum of vegetation, and is quite firm. It should however, have a fine layer of sand or dust.

School playgrounds, little league parks, and construction sites are good places to start looking for such a practice site.

Managing the Group, Spacing

If it is your intention to conduct this training for a large group of people you should select an area that allows each tracking team to move their tracks in the same direction with about a twelve to fifteen foot spacing between each team.

All the exercises in this book can be laid out and practiced by a single person working alone. However, the addition of a partner helps, and a team of three people is ideal. Many times over the past several years we have attempted to train huge groups with too few field instructors. This has forced us to form tracking teams with far more than three members. It can be done but the quality of the learning experience diminishes greatly with each additional person beyond the optimum of three. Therefore, if you wish to train thirty people, you should divide them into ten three-man teams which will require a training site at least 120 feet wide in order to maintain the minimum safe spacing between groups.

If you are training a large group you should put a single person in charge of laying out all the sets of tracks. He should lay out the practice tracks to be followed by groups 1,3,5,7, and 9 while wearing one pair of shoes with distinctively marked soles, then change to another pair of shoes containing

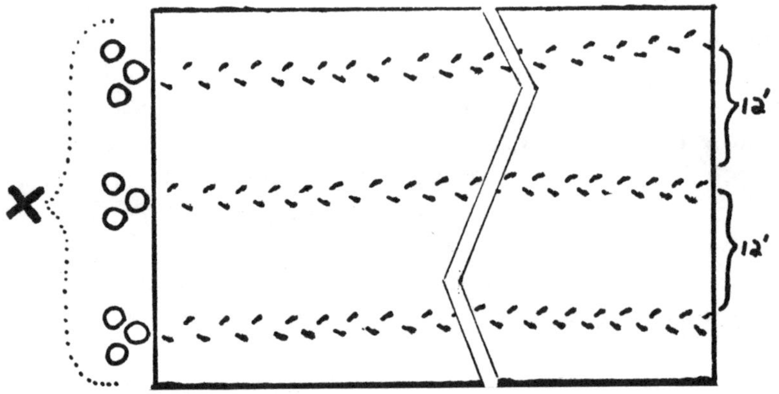

O -- STUDENT TRACKER X -- FIELD INSTRUCTOR

Above drawing illustrates layout of field exercise showing three sets of parallel tracks with a twelve foot separation between sets. Student trackers are formed into three 3-man teams each consisting of a point man with two flankers. A single field instructor supervises the progress of the three tracking teams.

uniquely different markings to lay out the tracks for groups 2, 4, 6, 8, and 10. The reason for differently marked soles is precautionary. It is very difficult for the track-layer to maintain the minimum spacing between sets of tracks until he has laid out these problems a number of times. He is very apt to waver in such a manner that two different trails almost converge. If the trails should join and the tracks are identical, the students will be unable to successfully complete the exercise.

If it is your intention to have fifteen feet of spacing between your teams, your track layer must put a thirty foot spacing between the first set he lays down (number 1) and the second set (number 3). He must, therefore, be very careful to select a point, across the field near the finish line, that is as close as possible to being thirty feet from the finish point of his previous set of tracks. A common error is to have good spacing at the starting points but spacing that is too wide at the finish line. This results in sets of tracks that are not parallel but flare outward so as to cause the higher numbered groups to be working tracks at angles that do not directly approach the sun.

Direction of the Tracks

In the beginning problem, tracks should always be worked towards the position of the sun (even on overcast days) and with the sun at as low an angle as possible so that the slight indentations of the track will cast shadows. In other words, your tracking problems should be held in early morning or late afternoon. At these times the sun is as close to the horizon as possible and affords the most favorable sun angle for viewing tracks.

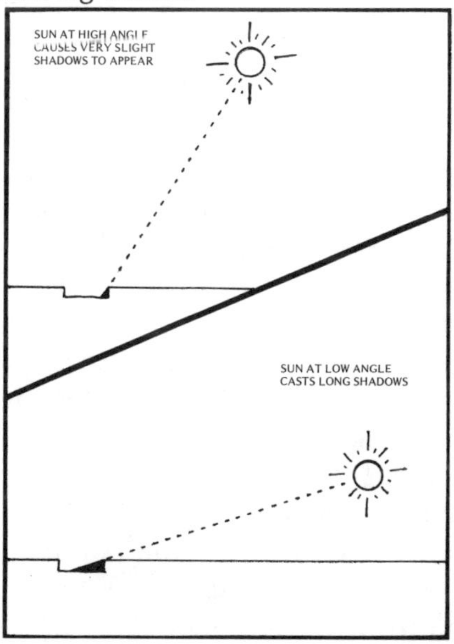

The single most important thing to learn about tracking at the outset is to utilize the sun. Try to always position yourself so that the track for which you are looking will appear directly between you and the sun. Caution and common sense must be employed, of course. You certainly cannot get atop a trail or you will obliterate it with your own tracks.

To learn the value of sun angle you should run your own experiment. Put down a track in the dust, then walk a complete circle around it, observing it from about five to eight feet away. The difference in your ability to see the track with the sun at your back as opposed to having the track directly between you and the sun is dramatic.

Also, run the same experiment within two hours of dawn, at mid-day, and again within two hours of sunset. The way the track disappears with the sun at its height and reappears in the evening will teach you a lesson about low sun angle that is far more graphic than I could accomplish with words.

Laying Out the Tracks

The track-layer begins laying out the first practice exercise by marking a straight line in the dirt across his path as a starting point. Just forward of the line he should make a circle in the dirt about eighteen (18) inches in diameter. He should scuff the area within the circle until it is loose enough to leave a distinct impression of his track, then placing his right foot within the circle (this being his first step) he walks normally in a straight line for about fifty (50) or seventy-five (75) yards (one hundred would be better, but it is usually difficult to find a field that long). At the end of his walk he should mark his finish line then circle away from the tracks and staying at least thirty feet from, and parallel to, his original route, return to the starting area. If the training is for a large group he must return by a route that gives wide berth to ALL the practice sets of tracks.

Identifying the Track

Once the tracks are laid out the training is ready to begin and each three-man tracking team should take a position at the start of a set of tracks. Once at the starting point as much time as necessary should be taken to study and

Group of student trackers at their starting points.

properly identify the track. Slight, subtle differences between very similar tracks have meant the difference between success and failure on a number of searches, as well as many law enforcement cases with which I have been involved over the years.

The track of your quarry must be examined carefully and often. Subtle cuts or markings may not appear in one track but may be discernible in another, therefore, identifying the track must be a continuous information gathering process.

Study the encircled track carefully. You should measure its length from the rear of the heel to the tip of the toe. Always when measuring tracks, be careful not to let your hands or the measuring tape touch the ground, thus making more sign. Avoid adding any more confusing marks to those already on the ground.

If there is a heel, measure its width and length. Measure the sole at its widest point. Note whether the toe is very pointed, moderately pointed, rounded, or blunt. Note any stitching or nail holes. Note any pattern or design.

Take a piece of paper and draw a picture of the track, putting in all marks and designs. If there are lines or bars across the track, separately record their exact number, coarseness, and approximate spacing. If you are working with a partner or as part of a tracking team, practice communicating the track description by stationing the other person, or persons, far enough away from the track so that they cannot see it and have them draw the track from your verbal instructions.

Tracker making a drawing of footprint.

Footprints are the best clues you can hope to find while searching for a lost person. They are the only certain evidence that a person is going to leave behind. They point the novice in a direction where the victim was once headed, and they lead an expert tracker directly to him. Being able to describe these clues to another search team, ten miles away, by radio, in descriptive terms that leave them with the same mental picture you possess cannot be overemphasized. I feel that track description is of such importance that I devote an entire chapter of this book to its coverage.

A sound general rule is to use descriptive comparisons that everyone can relate to, such as: "a dime sized circle" instead of a "circle that is about 11/16 inches in diameter" or "bars about the thickness of a wooden match stick" instead of "bars approximately 1/8 inch wide."

The value of proper identification can sometimes be of paramount importance:

> One of my most challenging tracking experiences began to unfold on the night of July 3rd, 1962, when a small Mexican national entered the United States surreptitiously near Tijuana, Mexico, and started a walk which he hoped would bring him to Los Angeles. We cut his track at dawn the following morning and started following it. With a team of trackers always on his trail and other officers cutting for sign ahead, we had, by late afternoon, taken the trail over 35 circuitous miles to the intersection of Main and Second Street in El Cajon. At this point our quarry turned west on Main Street and began walking directly through the center of the city which, at that time, had a population of nearly 40,000 people. However, the markings on the bottom of his shoes were so unique that we were able to find and identify his track in the few dirt areas available. We continued following him directly through the busiest part of the city for nearly two miles to its western edge, southwest towards the freeway, then back north on the railroad tracks where, at nearly nine o'clock in the evening we caught up with him and made the arrest.

Importance of Step Interval

Once you have found every idiosyncrasy of the right shoe of your practice tracks, looking particularly for the things that make it unique, like cuts or worn spots, you should look forward about eighteen (18) inches and to your left about eight (8) inches for the back edge of the heel of the next track. In very hard ground, the curved back edge of a heel is the mark you are most apt to see. Once the toe of one track is established in relation to the heel of the succeeding track you will have the information needed to determine your subject's "step interval".

The step interval, or the distance between tracks, is the information needed to transform your walking stick into a handy-dandy track finder.

When I was first learning to track, the more experienced tracker who was giving me tips, advised me to regulate my stride so as to match the stride of the person I was following. I was to do this by placing the tip of my right toe at the rear of my quarry's right heel and step forward adjusting my stride so that the tip of my left toe came to rest at the heel of my quarry's left heel. In theory, I should always find the track I sought just ahead of my foreward foot.

I tried this system diligently but found many faults with it. It was too easy to mis-step and obliterate the trail; I was too likely to blunder past a sudden turn, etc.

Through experimentation I began marking the distances between my quarry's tracks (the step interval) on a walking stick and holding it near the ground so that its tip would pinpoint the location of the next track.

It worked well for me, so I continued to use it, but I did not realize what a really effective tool it was until I began encouraging students to use it during the evolution of our step-by-step learning technique and our two-day tracking indoctrination course.

I have gradually come to realize that the utilization

of the tracking stick has probably been the greatest factor in helping non-trackers become acutely track conscious very quickly and is therefore one of the biggest reasons our instructional system has enjoyed such outstanding success.

Utilization of the Tracking Stick

The correct manner of establishing the marks on your stick is to start at a point where your subject was walking normally and the heel and toe marks of both shoes can be seen.

You start by placing the point of your stick even with the heel of the forward track, then moving your rubber band marker so that it is directly above the tip of the toe of the rear footprint. This gives you the correct step interval and is the piece of information of most importance to you.

Once this forward rubber band marker (toe marker) is in place you alter the angle of the stick so that it is directly over the center of the rear footprint. Placing your established toe-marker directly over the tip of the toe, you move your rear rubber band marker (heel marker) until it is directly over the rear edge of the heel of your subject's footprint.

You now have the length of your subject's foot shown on the stick (between the heel-marker and the toe-marker), the length of his stride (between the heel-marker and the tip of the stick), and the vital information: the step interval (from toe-marker to tip of stick).

With this information on your stick you are now able, upon finding the slightest trace of a footprint, to establish the approximate location of the tip of the toe of the found track and by pivoting your stick at this point (the toe-marker point) and swinging it in an arc, pinpoint the area where the succeeding track should appear. In swinging your stick, it is best to start with a small arc that initially allows the tip of the stick to travel no more than ten inches. If *careful* examination of this area reveals nothing, the length of the sweep can gradually be increased.

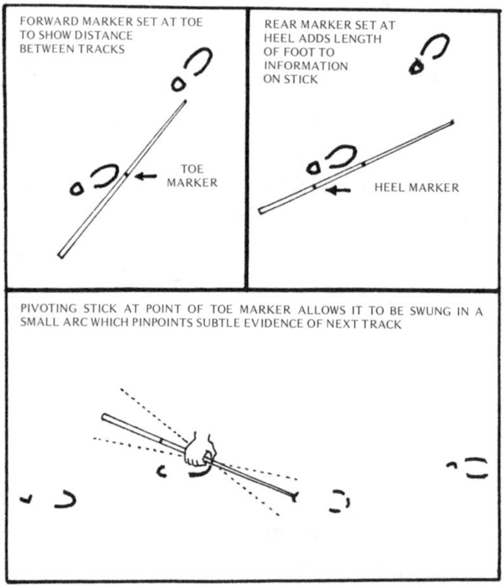

It should be mentioned at this point that a person's stride will vary, occasionally as much as six inches, even when they are walking on level ground. Therefore, in marking your stick you try to establish your quarry's *average* step interval. Mark your stick accordingly and do not change it.

Marking Tracks

To properly utilize the tracking stick it is important to always know whether you are looking for a right track or a left one. If you have just found a right track and forget what you are looking for, you may swing your stick to the right again causing you to be at least sixteen (16) inches off course. For that reason, we advocate making a curved mark behind the heel of each track to make its whereabouts obvious, and by putting a tail outside of the arc you can indicate whether it is a left or right track.

This has become the standard way of marking tracks and all search personnel who have been exposed to our step-by-step training will recognize it.

A fact well worth mentioning at this time is that entire footprints only appear when the ground is very soft. A complete, easily identifiable footprint only shows in dust and sand so easy to track through that your grandmother could do it. Expert trackers seldom see footprints but do their tracking by following partial tracks, flat spots, scuffs, and bent vegetation. These slight indicators are what the tracker calls "sign"

It is impossible to say how far a tracker might have to follow sign before he finds a full footprint that reveals the markings that positively identify his victim, but when these full tracks are found, they should be uniquely marked. Most of us at El Cajon draw a circle around the track. Some Mountain Rescue Association teams prefer to encircle it with rocks or trail tape. As long as you make it obvious, we cannot quibble with it, just mark it well.

Illustration shows tracks marked to indicate left or right prints. Tracks are circled when they reveal enough identifying marks to constitute a "positive make".

Realize that the footprints and sign which you are following will not be the only ones in the area. People, who have attended our two-day indoctrination course at El Cajon, or have heard my lectures at Search and Rescue Seminars should remember my emphasizing the point that the problem in finding a lost person is never the absense of sign, but the maddening excess of it.

Many times in searching for lost persons, or in tracking aliens and smugglers who have entered the United States illegally, the best of us have had another trail blend into the one we are following, then branch off. We may follow the wrong one for several hundred yards before we get a positive identification and discover it's the wrong track.

We must then go back to our last "certain" track (the one we circled) and start again, being careful not to take the wrong fork a second time. This is the reason we mark the positive tracks well and the reason we work better in teams of three.

The Three Man Tracking Team

When three of us are following a set of tracks and a conflicting trail joins the one we are following, we split and follow each. Whichever one of us discovers that he is on the correct trail advises the others to abandon their trail and regroup on the correct one. This is apt to be done many times on a trail, particularly if it is in an area which has had a great deal of foot traffic.

In our basic tracking classes we like to group three people on a single set of tracks. The designated leader of the group works as "point man", a position that puts him slightly forward of his two "flankers". It is his responsibility to follow the set of tracks and maintain order by seeing to it that neither flanker gets ahead of him and thereby in a position to obliterate the tracks. He must be particularly watchful for the "cheaters" in every beginning group who do not have the patience to look for the subtle tracks. These people are always looking ahead for an obvious track and keep trying to creep ahead for a better vantage point.

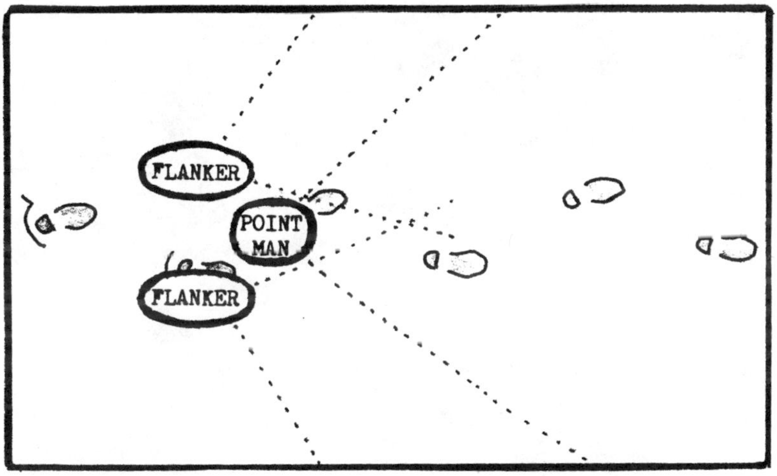

Illustration shows starting position for members of the three man tracking team and indicates their areas of viewing responsibility.

The first responsibility of the flankers is to watch to the side for another incoming trail and to be sure the trail the group is following does not make a sudden turn. Their secondary responsibility is to assist the point man in finding the next track. It is wise to rotate the flankers up to the point position. Working the point position is very fatiguing, so to keep the team fresh, each person should take his turn.

The major things accomplished by the three man tracking team for beginners is that it builds confidence, and lessens the chance of error. What three men agree upon is more apt to be correct than what one man decides by himself.

Tracking is done most efficiently as a team effort and we try to develop the appreciation of this fact in our training exercises.

Instructor indicating with pointer the correct pivot point (toe-marker point) of point man's tracking stick. Tracks proceeding to left and all trackers properly positioned so that tracks are directly between themselves and the sun.

What to Look For

You have now been told how to lay out the beginning problem, how to identify your victim's track, and how to utilize the tracking stick to find the approximate location of the next track. You have been told why you must mark the tracks either left or right and why you must remember which track you are looking for. You have also been told why you should undertake this exercise as part of a three man tracking team.

You have been told very little about precisely what to look for. This, of course, varies greatly, depending on terrain. In the beginning problem, you are supposed to be working on flat ground which is free of vegetation. Therefore, you should look for the curved rear edge of the heel, the curved point of the toe made evident by the propelling push of the walker, flat spots, small rocks pushed into the dirt, and small rocks uprooted from their natural beds and scuffed forward in the direction your subject went.

Of these clues the most important are flat spots. Only hooves and something man-made will leave flat spots. Small animals will scuff and move pebbles, but if there is a flat area as big as a nickel it has to have been made by a hooved animal or a man.

If the ground is so hard that you are not sure whether the flat spot was caused by a man or a hooved animal, test the ground to see whether your own weight will flatten it in a similar way. If not, then the flat spot had to have been made by something that exerts a lot more weight per square inch of foot area, hence a hooved animal.

Get your nose to the ground and look for anything!!!!!

Do not, under any circumstances skip a track! Just because you can see a big obvious track a little bit ahead, do not jump up to it! My half-blind, idiot Aunt Clementine can follow big obvious tracks. She has never learned a thing from them and neither will you.

The extremely difficult tracks, the subtle ones, the ones you have to hunt for half an hour are the only ones that teach you anything. When you are forced to dig for the most minute hint of a track you are on your way to becoming a tracker.

When we say step-by-step, that is exactly what we mean. Find every track, not 106 out of 107. There is no premium on speed in learning to track. The important thing is the self-discipline of making yourself stick with it to find the tough ones.

Keep in mind what it is that you are trying to accomplish. You are *not* trying to find a lost child in the wilderness. You are trying to learn how to track so that you may have the skill to find that child at some time in the future. It would be foolish to insist on finding every track in an actual

Students searching for extremely subtle tracks.

search when there are obvious ones a few feet ahead. It is just as stupid and foolish in the learning situation to follow only what you have always been able to follow and to skip over the magnificent lessons that will teach you what you need to know.

Realize, that in order for a person to become a skilled tracker, superior eyesight, though helpful, is not an absolute necessity. As in looking at material written in a foreign language, the trick is not in "seeing" it, but in interpreting what you see. A great deal of sign is seen by the rank beginner and disregarded as insignificant because he has not learned to glean its meaning.

You should now lay this book aside and read no further until you have gone into the field and practiced this beginning problem at least three times for a duration of from two to four hours on each occasion.

4

Utilizing Sun Angle

Phase I Exercises, Continued

In the preceding chapter you were presented with the most basic tracking exercise. It was designed to: (1) Teach you to carefully identify the track to be followed. (2) Utilize the tracking stick in order to find the most obscure tracks. (3) Mark tracks so that you remember whether you are looking for a right or left track. (4) Learn the effectiveness and confidence building value of the three-man tracking team. (5) Develop the good habit of looking for every track. (6) Break the bad habit of looking only for obvious tracks. (7) Learn some of the things that constitute foot-clues (sign).

If you followed instructions and practiced the beginning exercise diligently you probably amazed yourself at the amount of sign you were able to see. Probably you had no idea how much visible evidence you have been ignoring and stumbling over (and obliterating) for all these years and if you learn only this and proceed no further towards your goal of becoming a tracker I will feel satisfied that my effort has been worthwhile.

The next three exercises are designed to: (1) Reinforce and refine what you learned in the first exercise. (2) Develop good habits through repetition. (3) Teach you to utilize the correct sun angle.

Exercises 1 through 4 present the students with a solid foundation for learning to become a tracker and constitutes Phase I of your training. These exercises present the basics and have come to be known as our step-by-step teaching method. Since there were so many people involved in police work and search efforts who needed to be introduced to these basics, we have focused all of our energies over the years on teaching only to this level of beginning student. The result has been that we have whetted a lot of appetites and made large numbers of students track aware but have not provided them with the guidelines to proceed to the limit of their capabilities. The Phase II and Phase III sections devised for this book should adequately fill the needs of these serious tracking students.

This Phase I training in track awareness, the step-by-step method, is something I have been involved in developing and teaching for about ten years. This standard two-day course consists largely of salty stories, lies, tales of derring-do, a slide show, and two half-day field exercises. In addition to this I have written several articles on tracking, the section on tracking for a manual on law enforcement, and the course on tracking that is to be included in the curriculum at our new U.S. Border Patrol Academy. Throughout all of this I extol the virtues of our step-by-step technique of teaching tracking. My purported purpose for all of this is supposedly to turn out legions of skilled trackers and surely, to some degree, this will be accomplished. However, this is not my real purpose;

my strong underlying crusade is to cause every person who enters a search area to become acutely clue conscious so that they will not do more harm than good.

Every person who enters a search area is a potential clue destroyer and as such is a potential killer.

As a tracker, I know from much experience how often a good tracker can solve the mystery if he is not preceded into an area by a group of well meaning clue destroyers.

Since track consciousness is clue consciousness, since tracks are the only clues that you can be certain a lost person will leave behind, and since they will be by far the most numerous of the clues left behind, it is arrogant of a person to believe he is a competent searcher if he has made no effort to acquire some skill at finding tracks.

The exercises outlined in this and the preceding chapter are designed to quickly teach you how to see and interpret sign. However, we have discovered that it goes beyond that and creates in you an awareness that carries over into many other things. It actually greatly increases a person's powers of observation and if there was no other reason than this one, I feel it is sufficient justification for a person, involved in search and rescue work or law enforcement, to invest the time and energy necessary to learn something of the age old art of following sign.

Definitions and Terminology

There are several definitions that should be learned in order to follow a tracker as he fondly discusses his art. A number of terms that we are going to be using in this book sound similar but actually describe operations that are uniquely different.

First you should realize that "signcutting" and "tracking" are distinctly different operations which depend on the finding of "sign". "Jump-tracking" is neither signcutting nor tracking but is a novice's inept attempt at both;

Clues

Clue Destroyer

and "step-by-step tracking" is more than anything else, a teaching technique.

Exploring a dictionary will do very little to help you understand what "sign" means to a tracker. You will find some descriptive words such as "mark", "trace", and "vestige", which do apply, but what a tracker means is: "sign is any evidence of change from the natural state that is inflicted on an area by a person's passage."

A careful scrutiny of the dictionary will gain you nothing in your search for the word "signcutting". It is a pure word in that it has meaning only to practitioners of the tracker's art. It simply means: "looking for sign in order to establish a starting point from which to begin tracking."

Tracking means: "following someone, or something, by stringing together a continuous chain of their sign." "Step-by-step tracking" is a disciplined teaching technique wherein the tracker sees each step in sequence and proceeds no further than the last visible track. It is a standard that an expert will strive for at all times, but in actual tracking operations is not always possible to attain. In an actual tracking situation an expert tracker tries to find some visible clue at each point where his quarry might have stepped. However, under arduous circumstances, even the best tracker must accept small gaps in his continuous chain of evidence. The key phrase is "small gaps".

"Jump-tracking" means: "finding a big obvious footprint, then proceeding along the indicated route of travel until another obvious track is found." The distance between these tracks is often extreme, fifty feet is common, several hundred yards is not uncommon. A great deal of guess-work is involved, a large portion of luck is needed, and practically no skill is required since the tracks found are of the obvious type that almost anyone is capable of seeing.

In signcutting you are not trying to follow tracks as you do in tracking, but you are trying to find just one or two very visible tracks as you do in jump-tracking. The principal difference between the two being that the process of sign-cutting should NEVER be done by walking the same direction that your subject would be walking. Signcutting is done by walking a path that would intersect the path of your subject at approximately right angles. It is done in an area where there is little chance that the tracks of the "cutter" could be confused with those of the subject, and it is done by picking terrain where sign should be easily seen.

I have several dear friends in search and rescue work who, though they are not trackers, have gone to great lengths to popularize tracking as a rescue tool. They felt uneasy with the word "tracking", so by fussing with it a little they came up with the more descriptive term "man-tracking". It made me feel a little uncomfortable because it is not a term a tracker would use, but I let it slide. Then as more amateurs were attracted to the learning of the art I began to hear terms like "signs of man", "signs of track", or simply "signs".

None of these are genuine terms. They are not the words or phrases that a real tracker would utter. I have had the privilege of tracking and talking tracking with grizzled mountain men from the High Sierras to the Great Smokies and have been pleasantly surprised to find we speak the same language. Tracking is an eons-old art and entitled to the dignity of its own terminology. What I do is "tracking" and I do it by following "sign". In this book I intend to make every effort to avoid the vulgarizing of tracking's colorful and traditional terminology.

Phase I Exercises with Changing Sun Angle

In all Phase I exercises you should lay out your practice set of tracks in a straight line on level ground. This allows the walker to move at a steady, even gait that produces a relatively constant stride. Knowing the walker's stride and the distance between his tracks allows you to mark a stick with these distances. When this stick is held near the ground and properly utilized it will usually pinpoint the location of the next track.

The site of Phase I exercises should be chosen so as to be largely free of vegetation, heavy rock outcropping, and miscellaneous ground cover; the purpose being to eliminate as much of nature's camouflage as possible. The subtle evidence that a human foot leaves on the surface of the ground is hard enough for the beginning tracker to find without the additional distractions of ground cover.

The tracks in the first exercise were aimed directly towards the position of the sun because tracks are most easily seen when they appear directly between the viewer and the light source.

Lastly, the initial exercise was to be run very early in the morning or very late in the afternoon so that the sun would be at a low angle. That was because light which approaches the ground at a low angle causes an accentuation of shadows that makes all indentations in the ground easier to see.

If you worked at the beginning exercise on at least three occasions and devoted as much as two to four hours to each outing, you should have established some good habits, sharpened your eye, and become comfortable with the tracking stick. You should now be ready to deal with the problem of changing sun angle. Exercises 2 through 4 alter nothing except the direction of the practice tracks in relation to the sun. Before going further the complete Phase I exercises should be listed; they are:

Exercise	*Slope*	*Vegetation*	*Time of Day*	*Direction of Tracks*
1.	Flat	None	Early morning or late afternoon	Directly towards sun

Exercise	*Slope*	*Vegetation*	*Time of Day*	*Direction of Tracks*
2.	Flat	None	Early morning or late afternoon	45 degrees to 90 degrees away from sun
3.	Flat	None	Early morning or late afternoon	90 degrees to 135 degrees away from sun
4.	Flat	None	Early morning or late afternoon	135 degrees to 180 degrees away from sun

In the preceding chapter you were encouraged to work on Exercise 1 at least three times before going further. You should now do Exercise 2 three times before moving to Exercise 3 and continue that pattern until each exercise has been done three times. In doing these exercises you should increase the angle slightly each time, in other words Exercise 2a at an angle near 45 degrees, Exercise 2b at an angle near 60 degrees, and Exercise 2c at an angle that approaches 90 degrees.

In this group of exercises you will learn to position yourself so that the track you are seeking will appear between you and the light source. This is extremely important in tracking. On hard surfaces correct sun angle often makes the difference between seeing a track or not seeing it. It does not matter if the sun is obscured by fog or dense clouds, the track will always be easier to see when it is directly between you and the true position of the sun.

You will learn to use caution in these exercises. It is obvious that you cannot position yourself wherever you wish or you will end up atop the tracks and obliterate them. It is possible, however, to stand to the side of the tracks and look

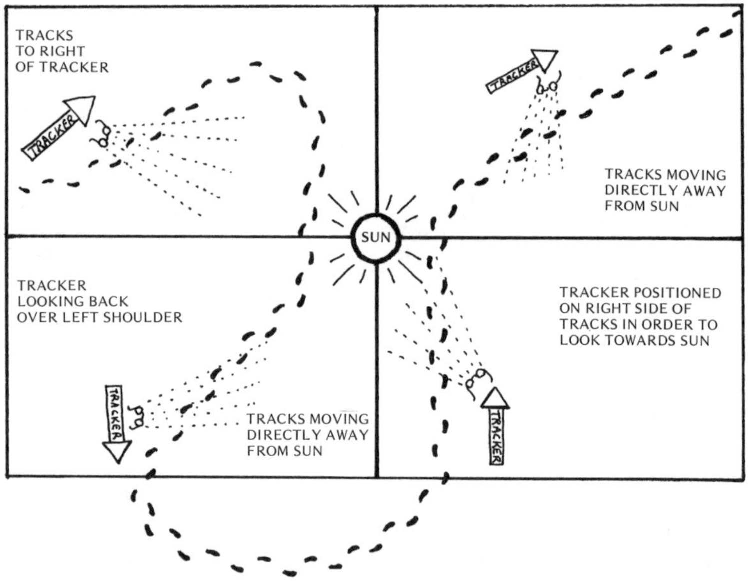

across them towards the sun. If you are attempting to follow tracks that run 180 degrees away from the sun (as will be the case in Exercise 4c) you should stand to the side of the trail and look back over your shoulder. This closely approaches the desired angle for efficient viewing.

Exercise Variation

In the foregoing exercises (2 through 4) you will be doing nine field exercises for from two to four hours each. If the training site you have selected is too easy you will find yourself being bored by the lack of challenge. Do not use this as an excuse for skipping parts of the training. Repetition is absolutely necessary for developing good habits. Try some of the following tricks to bring the difficulty of the problems up to a challenging level.

1. Have the track-layer wear a flat-soled shoe that is devoid of distinctive marks. This robs you of the lines and geometrical shapes that catch

Tracker following tracks with head turned to best viewing angle.

the eye.

2. Have a light person such as an eighty pound child lay the tracks. The reduction in weight reduces the damage inflicted on the ground.
3. Have the person laying the tracks wear a very soft shoe such as a moccasin. Sign is caused by a soft surface being marked as a result of colliding with a harder surface. Usually the weight of an adult's body propelling the hard surface of a shoe into the ground produces an adequate amount of sign. However, if the shoe is very soft it will do some of the yielding, thereby reducing the amount of visible damage inflicted on the ground.

Motivation

You will notice that this book has so far been surprisingly devoid of brilliant inside tips that will turn you into an instant expert. It contains not a single "smart pill" that will allow you to acquire the knowledge without doing the work. As mentioned before, it takes a solid groundwork of fundamentals and *plenty of practice* to be an accomplished tracker.

Let's hope you are still motivated to give it a try. But if it sounds like a lot of hard work, and you rationalize your laziness by telling yourself that you never wanted to be a tracker anyway, let me leave you with this one thought. We need you in search and rescue work, but if you are going to come into a search area with your big feet and your big vehicle tires, destroying all those clues we trackers need to save a life, I would prefer that you stay the hell away! Too many times we have followed a set of tracks up onto a dirt road only to see them disappear under a ton of wide-tired jeep tracks, and too many times we have followed a set of tracks down into a drainage to find it go under the tracks of a swath of line searchers.

Too many times? If it had only happened once that would be too many times.

We are all human beings and we make mistakes, but if people should die as a result of our mistakes, we have some responsibility to our fellow man to put forth the effort needed to prevent those mistakes from being repeated.

5

Sign

What to Look For

When a rank beginner hears someone mention tracking, he immediately envisions a person following footprints. It should be obvious by now that if there were complete, visible footprints in evidence, tracking would be no problem at all. To become a tracker you must learn to follow sign.

As mentioned before, sign can be defined as any evidence of change from the natural state that is inflicted on an area by a person's passage. This evidence can be lumped into two broad categories, Conclusive Evidence and Substantiating Evidence.

Conclusive Evidence is a bit of sign which, standing by itself, with nothing else to support it, can conclusively be said to have been caused by a human being.

Substantiating Evidence is a disturbance which could

easily be caused by something other than a human being and must be found in combination with several bits of such evidence before it can reasonably be determined to have been caused by man.

Since the particular things that constitute sign vary greatly in different types of terrain and since the specifics of what you look for in a Mexican desert are of little value to a student in the Rocky Mountains, I have felt the need, for several years, of finding a way of generalizing about sign. For about three years now I have been trying to analyze what my mind and eye are searching for when looking for sign. After much inward scuffling, I have come to realize, that regardless of terrain, there are four basic things that catch my eye. These characteristics of sign are:

1. Flattening — A leveling of rocks, twigs, leaves, or dirt caused by the flat sole of a shoe pressing the ground under a person's weight.
2. Regularity — An effect caused by straight lines, circles, or other geometrical shapes being pressed into the ground leaving marks that are not normally found in nature.
3. Color Change — A difference in color or texture from the area that surrounds it.
4. Disturbance — Evidence of recent change or rearrangement.

Let us now look at some examples of sign, and place them in their proper category.

CONCLUSIVE EVIDENCE

Flat Spots (Flattening)

Of all the faint disturbances that may be visible on hard ground, nothing says "man" more positively

Flattening

Regularity

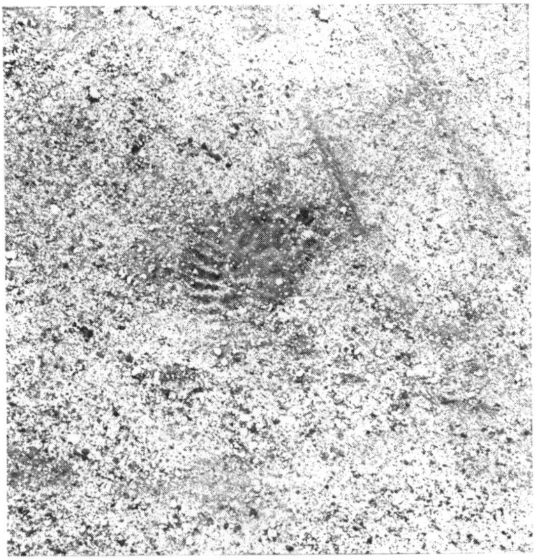

Color Change

Disturbance

than flat spots. Animals, birds, reptiles, insects, and wind blown objects are continuously making a profusion of marks on the ground that add confusion to the task of the tracker, but only man made objects (shoes, tires, rifle butts, etc.) and hooved animals can make flat spots.

When a flat spot no larger than a nickel is seen on the ground it may not be immediately discernible whether it was caused by a hoof or a shoe, but a simple testing of the ground with your own foot will give you the answer.

All hooved animals exert considerably more weight per square inch of foot surface than does man. Therefore, by placing your own foot near the flat spot in a *normal walking motion* (do not stand on one foot and stamp or twist the testing foot) you will reveal whether you leave a similar mark. If the ground is so hard that you left nothing as flattened as the flat spot you found, then it may be presumed to have been caused by something other than the person you are seeking.

Buried Pebbles and Small Sticks (Flattening)

When pebbles or small sticks are pushed below their natural bed it generally means something very hard, such as shoes or hooves have stepped on them. Since animal paws are usually softer than the surface of the ground, they will do the "giving" when contact is made with pebbles or small sticks.

If the ground is soft enough to allow paws to impress these hard objects into the ground, it will be soft enough to reveal the whole outline of a shoe or hoof. Again, check to see how much of your own track is revealed.

Rear Curve of Heel (Regularity)

The rear curve of a heel is probably the piece of conclusive evidence that occurs most commonly. In many instances, it will be the only bit of distinctively human evidence a foot will leave on hard ground. This is because in

Flattened area with buried pebbles

normal walking, on level ground, the entire weight of the body is transferred forward with each step and focused, at initial impact with the ground, on the small rear edge of the heel.

If an indentation is going to be inflicted on the ground at all, it is most apt to occur by this striking force. The uniformly curved indentation that results from this impact is very noticeable to a person with very little tracking experience. By the time a person has about 40 hours of tracking behind him he will be spotting this indicator very quickly.

Tip of Toe, Toe Digs (Regularity)

The weight of the body being shifted forward coupled with the "pushing-off" action of the propelling foot causes a digging action that leaves a distinctively uniform indentation in the ground.

The only problem with the "toe dig" is that it is not

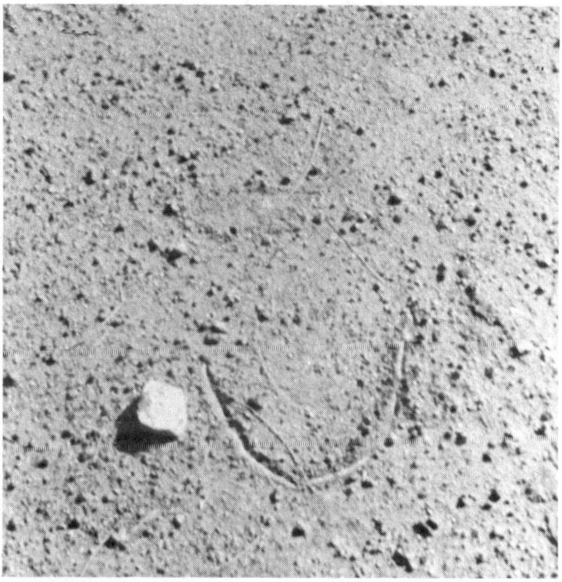

Curved line drawn in dirt outlines barely discernible rear curve of heel. Above the left side of the heel is a flattened area containing an impressed pebble.

as noticeable as the two preceding items of sign and it seems to take several hundred hours of tracking experience before the eye becomes trained to spot it readily.

It is one of those obvious clues which non-trackers, and beginning trackers see clearly, but lacking experience, fail to interpret correctly and thus disregard as being unimportant.

Bent Grass (Flattening, Color Change)

Human beings and hooved animals push grass downward in a similar manner, but there are several ways of telling the difference.

Since a man's foot is longer than that of a hooved animal, the length of the pushed-down area will be longer. The width of the pushed-down area will be widest for cattle and horses, somewhat less for man, and narrowest for deer, sheep, goats, etc. The most telling clue, however, is that the

Toe dig in lower right of photo, heel impression upper left.

sharp frontal edge of hooves tend to crimp the grass towards the middle of the pushed-down area while the smooth sole of a man's foot will not.

Bruised Stems and Stalks (Color, Flattening)

Stems and stalks of green vegetation which have been stepped on or kicked may return to their upright position in less than 24 hours, but the bruises inflicted may be visible, upon close examination, for several days. Often these bruises will be slightly flattened, telling you positively that the bruise was caused by man. Small animals seldom bruise vegetation at all but certainly cannot flatten it. Hooves which strike with enough force to bruise will nearly always cut.

Picking Up Mud (Color Change, Regularity)

"Picking up mud" is a tracking term that does not mean the walker periodically scoops up a handful of mud to

Flattened bruise mark on stem.

Close up of flattened bruise.

amuse himself. Mud tracking is too simple to dwell on, except in those cases where the walker is picking up mud on his shoes. In this case we are presented with a fairly difficult problem. What happens is that the walker enters an area where the soil has a gummy, claylike consistancy that sticks to the shoe. At the outset the walker will try to stamp it off or periodically clean it off, but he soon realizes the futility of this endeavor due to the fact that new mud starts adhearing to his shoe immediately. The walker soon resigns himself to this nuisance and continues walking until he has a large pillow of mud attached to each foot.

These blobs do not resemble footprints but they do leave their marks at regular stride distances and in a normal left-right walking sequence. The marks that are left on the ground alternate between tiny patches of surface dirt that have been removed (picking up mud) to chunks of mud that have fallen off once the weight has increased beyond the mud's adhesive power to hold it to the shoe. On these droppings you will often find the flat and uniform characteristics of the wearer's shoe.

Shine (Color Change, Flattening)

Shine is the most subtle bit of conclusive evidence, and unlike any other type of sign except grass trails, is far easier to see at some distance than up close. In dirt, shine is caused by the almost imperceptible flattening of dirt particles by the foot which creates a reflective surface that is only visible with an oblique light angle. The angle at which the light hits the tracks and is reflected into your eye is all important. That is why six or seven subtle footprints may be seen at a distance of thirty feet while you are unable to see them immediately in front of you, or beyond that point where the sun angle reveals them. Only by approaching these tracks will the succeeding ones be revealed. The same phenomenon occurs with less obvious grass trails and with trails

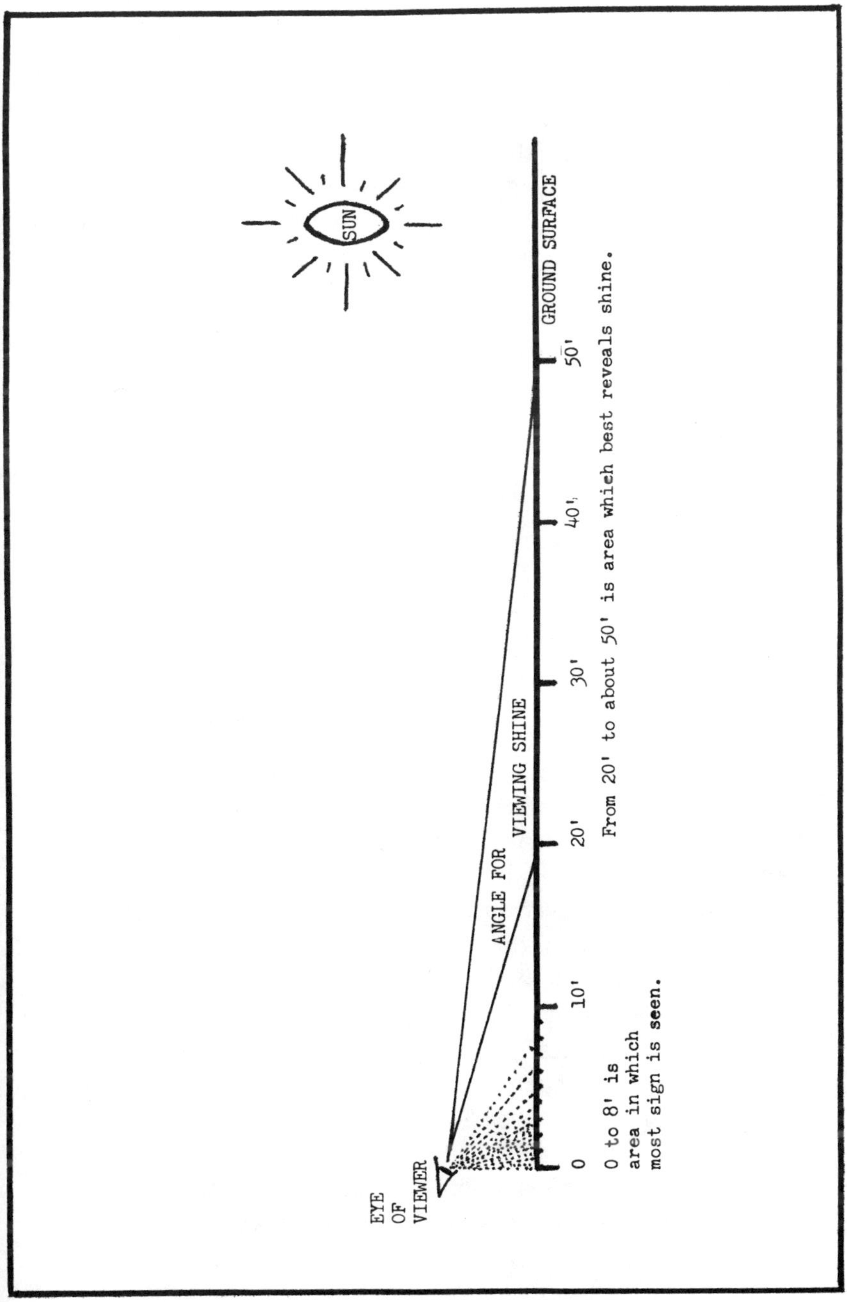
SUN
GROUND SURFACE
ANGLE FOR
VIEWING SHINE
EYE
OF
VIEWER
0
10'
20'
30'
40'
50'
0 to 8' is
area in which
most sign is seen.
From 20' to about 50' is area which best reveals shine.

across dead dry grass. Looking closely at the ground immediately in front of you does not produce the necessary light angle so you may have to look ahead (in the direction of the sun) for distances as much as several hundred feet.

In actual situations I have stood on hillsides and seen grass trails through a meadow a half mile away which were revealed by such a subtle change of color that they were absolutely impossible to discern when viewed at a distance of less than one hundred (100) feet.

As a general rule the most difficult tracking you will ever encounter will be on flat ground and this is the area that affords you the greatest opportunity to look ahead at great distances for that precise sun angle that will reveal "shine".

Under these circumstances you need to be very cognizant of the fact that the surest way to see sign is not always when it's one foot from your nose. When tracking gets very difficult it is very wise to look at great distances. Sometimes you will be shocked to see how obvious the trail is when viewed from the correct angle.

Transfer (Color Change)

When a person walks from one type of terrain to another there will often occur a few steps when material from one area is transferred to the other. When a person has been walking through mud and then enters either a grassy, rock covered, or paved area he will usually clean the mud off his feet immediately thus giving you at least a few steps to indicate the direction he has chosen to travel.

Similarly, if a person has been walking on a dusty trail he will transfer a few dusty prints onto a flat paved area. Rarely will such transfer be visible on the uneven surface of rocks and even if it does it would have to occur in an area so dusty that tracking would be very easy and you would have little need to be so carefully scrutinizing rocks.

Water also can be transferred to rocks or soil but will

Transfer of mud onto paved road.

only be visible on an extremely fresh trail. I have never encountered a situation where water transfer remained visible for as much as a day.

Animal Droppings, Human Feces

I suspect that all those reading this book have checked enough times in their lives to see if the toilet was flushing properly to be quite familiar with the sizes, shapes, and colors of human feces. Sizes and shapes of human feces do not vary greatly and can be easily differentiated from animal droppings of the same approximate shapes by the fact that there is no hair in them. It may be necessary to probe the feces with a stick in order to make this determination, but I strongly recommend that you lay aside any squeamishness you may have and never pass up the opportunity to probe this bit of sign. A great deal of information can be gathered regarding the diet of the defecator. Very black feces can mean a high degree of iron in the diet and as the color ranges to more yellow the more likely the person is to be slightly anemic. Roughage items in the diet tend to pass through the digestive system relatively unchanged allowing for things like the pulpy part of oranges to be easily recognized.

Pimento and red chili peppers are a very common part of the Mexican diet and are capable of passing through the body without losing their brilliant red color. Many times over the years I have, while in the border area, probed human droppings of questionable origin and discovered these telltale signs of a Mexican diet, thus causing me to follow the culprit in order to question him concerning his right to be in the United States.

All excrement, both animal and human, is warm and soft when it leaves the body and if stepped on by even the tiniest animals while in this soft state will squash and flatten. As it dries out and hardens greater weight is required to squash it. For this reason it is very unwise to draw conclusions from the fact that rabbit feces, deer feces, or other types of animal excrement have been flattened unless we have additional sign to go with it.

The age of human feces can be determined with about as much accuracy as can vegetation due to the fact that bacteria begins attacking it the minute it is exposed to the air and flies are attracted to it shortly thereafter. As I pointed out before, the color of feces is determined by diet but the *difference* in color between the darkened exterior crust and the interior of the feces is determined by heat and by bacteria action.

Interpreting this difference in color can be very tricky since we are greatly influenced by sharp contrast. Human feces resulting from a high iron intake diet will be very black and will not contrast much in color from the exterior crust even after several days, while the interior of more yellow feces will contrast sharply with its crust after only a few hours.

I prefer to determine the age of feces by relying primarily on the presence of flies and secondarily on the thickness of the crust.

Feces that are four hours old or less may have four or five flies attracted to it. From this point in time up to 24 hours it will likely only have one or two flies. Beyond 24 hours it will probably not have any. So in order to estimate how old it is, beyond 24 hours, requires that you probe it with a stick to determine its softness (moistness) and the thickness of its crust.

I suppose before leaving the subject of human elimination, it should be mentioned that the color of urine also reveals a great deal about the physical condition of the person urinating. The darker yellow the urine becomes, the higher the probability is that the fluid level in the person's body is diminished and that he is experiencing a degree of hypohydration. This difference in color is only really very noticeable when tracking a person through snow, but is very valuable nonetheless. As your quarry's urine becomes more orange the less time you have to find him in a healthy state.

Discardables

In the preceding chapter I touched briefly on the fact that discarded material does not occur frequently enough that time should be spent in teaching people to look for it. However, this does not mean it should be ignored. It is just that looking for the types of sign I have outlined in this chapter will hone your powers of observation much more sharply and allow you to spot these discarded items more quickly.

As tracking is an information gathering process, so is interrogation, and in order to know what types of discarded material might have belonged to the person you are looking for you should, when possible, delay starting your search until you have had time to talk to the lost person's family or companions concerning what the lost person might have been carrying.

As thorough an inventory as possible should be made as to the type and color of clothing, type of food, brand of cigarettes, etc. Determine if the subject chewed gum, ate sunflower seeds, compulsively chewed on a toothpick, or had any other personal habits that might provide you with a piece of litter that might be distinctively theirs.

One very important search I was on several years ago pivoted on the fact, which we had uncovered during preliminary interrogation, that the young man we sought was carrying dark blue, twelve gauge shotgun shells. We followed his tracks into a canyon containing nothing but huge rock boulders and were able to maintain the continuity of his trail by occasionally finding a few of these spent casings.

SUBSTANTIATING EVIDENCE

While an item of Conclusive Evidence can sometimes be found by an absolute non-tracker, and a great deal of it is found by beginning trackers, both of these groups are nearly completely oblivious to the abundance of Substantiating Evidence that almost leaps to the attention of an expert.

Bits of substantiating evidence are inconclusive when standing alone. However, by plotting their occurrence in relation to other bits of such evidence you can many times string together a continuity of sign that becomes conclusive.

Dislodged Twigs (Disturbance)

When twigs, bits of straw or small sticks have lain on the ground for any length of time, their outline becomes imprinted in the dirt. When something moves these twigs the outline or "bed" remains. Since the normal movement of very small animals can also dislodge these twigs, they are only indicative of human traffic when combined with some other form of substantiating evidence.

Dislodged Pebbles and Rocks (Disturbance)

Pebbles have beds just as do twigs, but moving them from their beds requires more pressure than it does for twigs. The direction twigs and pebbles have been moved indicates the direction of travel of the walker. Therefore, if several small pebbles in an area as large as a silver dollar have all been moved the same direction, it was likely done by a human foot. However, if several pebbles have all been moved the same direction in an area as small as a twenty-five cent piece, you would need some additional corroborating evidence, such as a dislodged twig within six to eight inches of the pebbles, and along the same line of travel, before you would dare assume that it was done by something other than an animal.

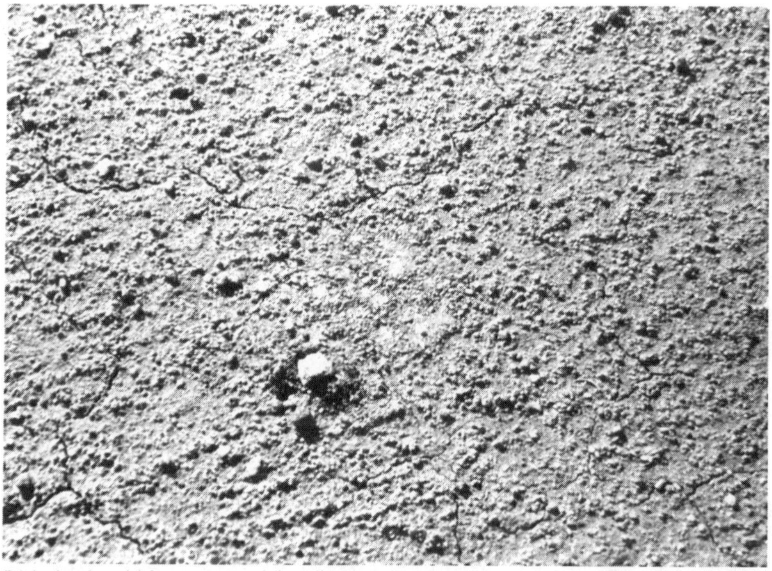

Dislodged pebble supported by flattened areas above and to the right. Combination of the two constitutes conclusive evidence.

Rocks up to the size of baseballs are similarly kicked from their beds leaving sockets that are quite prominent. However, when not accompanied by other corroborating evidence it only tells us it was done by a force equal to man's.

Fallen Petals and Leaves (Disturbance)

Fresh, pliable petals and leaves usually have to be knocked loose in order to become separated from the plant on which they grew. Their appearance on the ground strongly suggests damage done by a large animal or man. In this same category are tiny pieces of limbs which have fallen to the ground and which still have live, healthy greenery still visible on them.

Breaks in Fallen Branches and Twigs (Disturbance)

Branches or twigs lying flush upon the ground will likely only be broken by shoes or hooves. Hooves tend to split and splinter. These are very strong clues as to what caused the break, but in themselves are not quite conclusive enough to be listed under Conclusive Evidence, and so are relegated to the category of Substantiating Evidence.

Breaks in Branches and Twigs Above Ground (Disturbance)

In the preceding paragraph we described breaks in branches and twigs (tiny branches) that had been already broken from the trees and bushes upon which they had grown and were lying upon the ground. In this paragraph we are interested in the break that occurs above the ground when something brushes against branches and twigs with sufficient force to break them, but still leave them attached. The word "branch" is used in this instance to indicate a limb that is at least a quarter of an inch in diameter at the point of the break, while a twig will refer to all sizes of limbs smaller than that.

The force required to break a branch normally requires

the size and weight of an animal at least as large as a man, so it is a better indicator than is a broken twig. If close scrutiny of the ground in the area below a broken branch does not reveal a hoof print, then a strong case is made for the break to have been caused by a man.

Broken twigs on the other hand tell you only that some unknown thing brushed against the bush unless they are coupled with some other type of substantiating evidence.

If the twigs are broken quite high above the ground, in other words, higher than you would reasonably expect for them to have been broken by a large canine, then they must have been broken by a tall animal. So again, checking the ground below such breaks for the presence of hoof marks will give you a very strong indication whether the break was caused by a human being.

Leafy Surface Plants (Color Change, Disturbance)

In practically all geographical areas there appear broad leafed plants that grow very close to the ground. In scrutinizing an area where these occur you should look for one of the leaves on the plant, or possibly only the tip of a leaf, to be flopped over to reveal its underside. The more unnatural the position of the leaf the shorter the period of time has been since it was disturbed. The angle at which the leaf is canted indicates the amount of time that has passed since the leaf was moved from its normal position. In other words, the entire underside of the leaf does not have to be visible for it to be valid sign. If the leaves normally grow parallel to the ground, or with their tips slightly drooping, then a leaf whose tip is only elevated as much as 45 degrees should arouse your suspicions that it was moved by a force the plant does not normally encounter. Its slightly elevated angle will tell you that a period of time has passed since this happened and that the leaf has almost returned to its normal position.

These broad leafed plants, when stepped on, often

show faint evidence of bruising, not so much from the surface of the shoe, but far more commonly from the tiny rocks and sand that are pressed into their underside. These sharply pointed pebbles puncture the leaf, showing a dark green wound when it is fresh, that turns to black as the perimeter of the wound dies, and eventually leaves a withered scar that is grey or light brown in color. When these leaves are viewed from the standing position only a speckled appearance is noticed. Usually the leaf will have to be examined closely and often turned over in order to inspect its underside.

Interlaced Vegetation (Disturbance)

Many types of living things from small bushes to tall trees grow so close together that their limbs intrude into the same space to touch or overlap. When something large moves between them it pushes these limbs out of the way, then

Interlaced vegetation.

releases them causing a shuffling and catching effect that does not allow them to return to their normal position. Thus they will remain bent out of position until they are shuffled the other direction or until the wind shakes them loose. The sturdiness of the limbs involved and their height above the ground gives you a clue to the strength and size of the creature that pushed itself along that particular path.

Scuffed Lichen (Color Change, Disturbance)

Finding sign on solid rock is not always impossible. If the trail indicates that the person you are following walked onto a large rock simply proceed in the indicated direction and carefully inspect any growths of lichen that appear in your path. Do not be overly optimistic; lichen, like moss and pine needles is often cushiony, and by yielding, avoids having its surface broken. When stepped on, however, it is sandwiched between two very hard surfaces and in those places where its cushion is not thick it can easily be torn by the scuffing action of the shoe.

Dew Trails (Color Change)

When a tracker talks about a "dew trail" he is actually discussing an "absence-of-dew trail". In the jargon of tracking there are a number of such strange descriptive words and absolute misnomers; I will point out several others in the succeeding chapters.

Dew trails are sometimes noticed on short grassy lawns or across golf courses. They are darkened areas that reveal the outline of every step of the person who walked there. They are discernible because all the surrounding grass is heavily beaded with sparkling droplets of water while the stepped on area appears darker because of the *absence* of dew. While in short grass, dew trails are too obvious to even discuss in a chapter dealing with subtle sign, it should be pointed out that in tall weeds, wild oats, fields of wheat or barley, or areas with lush vegetation they are not nearly so obvious to an untrained eye, but are extremely helpful to the tracker, nonetheless.

Dew trail.

Dew, of course, is formed when moist air is cooled by direct contact with cold objects in an open area and it occurs best on nights that are still and clear. This condensation remains on the objects on which it is formed until it is knocked off or until the heat of the day evaporates it. This is a very good reason for starting your search for sign at the earliest light of day. In dense forests which have spongy moss floors or cushions or pine needles the only truly continuous trail you may find may be the ferns, leaves, and branches that have been brushed against causing them to lose their dew. All large animals will of course leave similar trails but you will be given a long continuous pathway which will surely at some time interval reveal whether it was caused by man or animal. Be sure and start early though, you will only have a few hours before sunlight will cause this moisture to vanish as magically as it appeared.

CONTINUITY

Continuity is of great importance in the evaluation of Substantiating Evidence. An indicator which, by itself, would prove nothing becomes almost certainly your quarry's track when it is sandwiched between two other bits of sign and is approximately a step interval from each. Utilization of your tracking stick as described in Chapter 3 will be invaluable to you as will your dedication to the principal of trying to find every track as we advocate in the step-by-step learning technique.

ALARM, OR MOVEMENT OF AN AREA'S INHABITANTS

One last area of visible sign which should at least be touched upon before terminating the chapter relates to curious behavior on the part of animals and birds that inhabit the area. Admittedly, in most instances, sudden movement or curious behavior is caused by the presence of a predator rather than by the presence of man. However, occasionally this behavior might be caused by man's presence so we must train ourselves to notice it and be curious about it.

Such evidence as the sudden flight of a flock of birds, the onset of squawking by geese or ducks, the sudden chattering of small animals, the abrupt silence of songbirds, the agitated fleeing of half-wild cattle or horses, the threatening swoops of hawks, or the ominous circling of crows and buzzards constitutes the type of sign to which the expert tracker is constantly attuned.

CONFIDENCE

People who are unsure of their abilities in whatever pursuit, seldom succeed, and while bravura is particularly inexcusable when lives are at stake, the honest, sincere confidence that comes from understanding what you are doing and having many times solved the knotty problem with which you must come to grips is absolutely necessary if you are to calmly, coolly, and methodically solve a difficult problem under extreme stress.

By going one step at a time you can know positively that you are still on your victim's trail. If you skip tracks you cannot be sure that subtle sign such as a broken twig was not caused by an animal, so doubts arise and your confidence in the trail you are following tapers off dramatically. You will most assuredly start disregarding good sign and eventually become hopelessly confused and distraught and will give up and quit.

Innumerable times I have seen learners give up and admit defeat when they were still right on their quarry's trail, but they had skipped so many tracks (jump-tracking) that they didn't KNOW it.

6

Slope and Ground Cover

Phase II Exercises

In the preceding field exercises you have worked on flat ground which was free of camouflaging ground cover. It would seem logical that if we were going to increase the difficulty of the problems gradually we might now alter one of these conditions, but certainly not both. What appears to be logical is not always so and this is one of those times.

We are going to alter both these conditions because that is the easiest approach to the problem.

If you have worked diligently at the preceding problems you should certainly know by now "how to look" for sign by utilizing the light source. You also should have learned a great deal about "what to look for". Specifically that means the curved rear edge of a heel, the straight front edge of a heel, the curved outer edge of a sole, the "dig" of a toe, the different color of a scuff, flat spots, pebbles impressed in the dirt, pebbles dislodged from their beds, etc. Generally speaking though, what you have learned to look for is any disturbance from the natural state.

Downhill tracking in area with varied ground cover.

As we move our tracking problems into areas that have vegetation, the specific things you have learned to look for will be very scarce. However, if you have conscientiously put in the hours of practice prescribed you will have developed the habit of generally looking for:

1. Flattening
2. Regularity
3. Color Change
4. Disturbance

A whole new "vocabulary" of sign must now be learned as you move to an area that is covered by vegetation, and the element that will save you from floundering is the fact that the problem will be laid out on a hillside.

Downhill Tracking

Tracks can either go downhill, uphill, or side-hill (staying at the same elevation while walking on sloping ground). Of these three situations, tracks are easiest to follow when they go downhill. In each of these three instances, the steeper the slope, the more sign the walker will leave.

In walking downhill people shorten their stride in order to keep the body upright without putting great strain on the legs. A downhill walker also transfers his weight to his heels in order to brake his descent. This, of course, causes a great deal of damage to the ground leaving much obvious sign.

If the person doing the downhill walking is wearing a shoe with a heel, rather than a flat soled shoe, the front edge of that heel will dig into the ground leaving a mark that is quite obvious and positively belongs to a human being.

The fact that people shorten their strides greatly in both uphill and downhill walking means that there will be many more foot placements on the ground, which, depending on the angle of the slope, could mean nearly twice as much visible sign.

When vegetation is added to the exercise, tracking becomes easier as the vegetation becomes more dense. Thick, high vegetation will leave a very obvious trail even on flat ground but when coupled with a steep downhill slope presents you with one of the easiest types of tracking problems.

We still must be concerned with marking tracks either left or right, but as we move our tracking problems away from bare dirt areas it will now become necessary to mark the tracks with trail tape.

In these Phase II exercises you should alter first the angle of slope and then the density of the vegetation in order to increase the difficulty of the problem. The downhill portion of Phase II training consists of the following nine exercises:

Tracker viewing obvious grass trail on slight downhill slope. Notice how lighter color of tracks contrasts with darker undisturbed area.

Exercise	*Downhill Slope*	*Vegetation*	*Time of Day*	*Direction*	*Difficulty*
1.	Steep	Heavy	Early or Late	Towards Sun	Most Easy
2.	Moder-ate	Heavy	Early or Late	Towards Sun	Very Easy
3.	Slight	Heavy	Early or Late	Towards Sun	Easy
4.	Steep	Average	Early or Late	Towards Sun	Easy
5.	Moder-ate	Average	Early or Late	Towards Sun	Some-what Easy
6.	Slight	Average	Early or Late	Towards Sun	Slightly Difficult
7.	Steep	Sparse	Early or Late	Towards Sun	Slightly Difficult
8.	Moder-ate	Sparse	Early or Late	Towards Sun	More Difficult
9.	Slight	Sparse	Early or Late	Towards Sun	Most Difficult

Each of these exercises should cover a distance of several hundred yards. You should take your time and attempt to find every step. If you finish in less than an hour and you have found every step, the problem was too easy for you. For an exercise to tax you and teach you, it should take about three hours. Alter conditions as necessary so as to present yourself with a problem that takes you about three hours to solve.

Sidehill Tracking

If you will think about it, you will realize that anytime you have ever walked on a sidehill you unconsciously have dug the uphill side of your foot into the dirt so as to create a tiny terrace that would provide you with solid footing.

The steeper the hillside, the more you have had to dig in your foot to maintain balance and the more damage you have inflicted. A trail left by someone walking on a sidehill leaves far more sign than will be left on flat ground but less than a person walking downhill. The specific things you should look for in sidehill tracking through vegetation are the flattening of those little terraces and the different color of downed vegetation.

As in the first nine exercises (Downhill Tracking) of these Phase II exercises, I change the problem by first altering the slope and then altering the density of the vegetation. However, you should notice that this does not necessarily increase the difficulty of the exercise. The second nine exercises (Sidehill Tracking) of Phase II are:

Exercise	*Sidehill Slope*	*Vegetation*	*Time of Day*	*Direction*	*Difficulty*
10.	Steep	Heavy	Early or Late	Towards Sun	Most Easy
11.	Moderate	Heavy	Early or Late	Towards Sun	Very Easy

Exercise	*Sidehill Slope*	*Vegetation*	*Time of Day*	*Direction*	*Difficulty*
12.	Slight	Heavy	Early or Late	Towards Sun	Easy
13.	Steep	Average	Early or Late	Towards Sun	Easy
14.	Moderate	Average	Early or Late	Towards Sun	Somewhat Difficult
15.	Slight	Average	Early or Late	Towards Sun	Difficult
16.	Steep	Sparse	Early or Late	Towards Sun	Easy
17.	Moderate	Sparse	Early or Late	Towards Sun	Somewhat Difficult
18.	Slight	Sparse	Early or Late	Towards Sun	Difficult

Uphill Tracking

When walking uphill, a person shifts his weight to the ball of the foot. If the incline is very steep only about half of his foot touches the ground. Since, under these circumstances, the heel will almost never touch the ground it can cause difficulties in differentiating human sign from animal sign. The more sandy and loose the soil on the bank, the greater the difficulty in distinguishing it as human sign. If you are fortunate enough to have firm ground, that telltale flattening that you have learned to look for will save you.

There will be plenty of disturbances in uphill walking due to the number of foot placements already referred to, and because of the unavoidable sliding that takes place even when wearing Vibram soled shoes or other shoes which have a high degree of traction. However, the thing that makes uphill tracking difficult is the lack of conclusiveness of the sign. There is much disturbance but very little of the regularity and flattening that positively identifies it as being human sign. Adding to the problem is the irregularity of the step interval and the fact the trail will waver considerably as the hill climber picks his easiest route of ascent.

As in the other Phase II exercises, these dealing with uphill walking begin by altering the angle of the slope and proceed to altering the density of the vegetation. This does not necessarily mean that the higher numbered exercises are the most difficult:

Exercise	*Uphill Slope*	*Vegetation*	*Time of Day*	*Direction*	*Difficulty*
19.	Steep	Heavy	Early or Late	Towards Sun	Somewhat Difficult
20.	Moderate	Heavy	Early or Late	Towards Sun	Somewhat Easy
21.	Slight	Heavy	Early or Late	Towards Sun	Easy
22.	Steep	Average	Early or Late	Towards Sun	Somewhat Easy
23.	Moderate	Average	Early or Late	Towards Sun	Somewhat Easy

Uphill walking. Notice slide area created by walker's previous step.

Exercise	*Uphill Slope*	*Vegetation*	*Time of Day*	*Direction*	*Difficulty*
24.	Slight	Average	Early or Late	Towards Sun	Somewhat Easy
25.	Steep	Sparse	Early or Late	Towards Sun	Easy
26.	Moderate	Sparse	Early or Late	Towards Sun	Somewhat Easy
27.	Slight	Sparse	Early or Late	Towards Sun	Somewhat Difficult

Continuing upward. Notice the inconclusiveness of the dented area where the left foot had been.

Degrees of Difficulty of Phase II Exercise

If you have been paying attention to the degree of difficulty of Phase II problems 1 through 27 you are now probably completely confused. In order to alleviate some of your confusion let us discuss what would happen to a single dandelion growing on a hillside when it is stepped on. Dandelions (and most vegetation) grow straight up. If a dandelion is growing on a 45 degree slope and is flattened by an uphill walker it is only pushed 45 degrees from the perpendicular. If it is flattened by a downhill walker it is pushed 135 degrees from the perpendicular. A sidehill walker would push it 90 degrees from the perpendicular.

The injury inflicted on the dandelion by the uphill walker is very slight and it will quickly recover. If there was much damage to the ground the rapid recovery of the dandelion (and other vegetation around it) would tend to hide this ground damage. Therefore, in this instance you would be better off if the vegetation was sparse.

The damage suffered by the dandelion as a result of being stepped on by a downhill walker is severe. It will take quite some time to stand back up. In this instance the more vegetation you have lying down the easier it is to follow the trail. Therefore, dense vegetation is desirable.

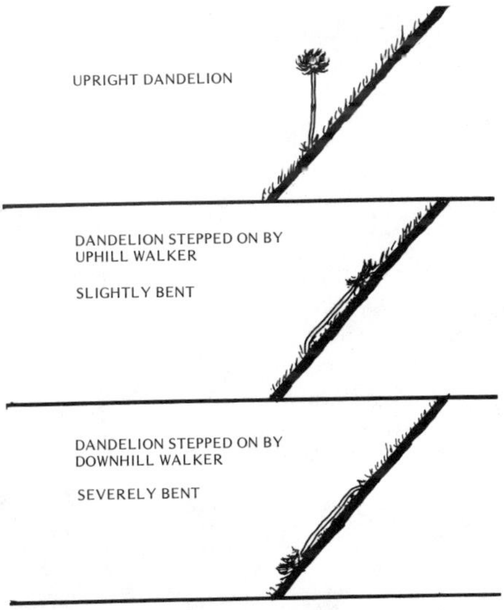

Briefly, that is why the altering of each condition will alter the degree of difficulty sometimes making the problem easier and sometimes making it more difficult.

Color Change

When a person is walking on sloping ground he will occasionally slip and slide. Usually these slides, scuffs and practically all other disturbances in dirt catch your eye because they are a different color from the rest of the terrain.

This difference in color is tremendously important in

Tracker closely examining vegetation

Closeup shows tracker discovering that top stem has been pressed into bottom stem with such force that bottom stem was dented.

tracking but difficult to discuss because it is almost impossible to generalize about it. I believe that slides and scuffs on a steep bank are usually darker than the surrounding dirt, but I have found some occasions when they were lighter.

The moisture content of the ground affects the color of a disturbed area. Usually an undisturbed surface is dryer and lighter in color than is the soil upturned from beneath the surface, but the make-up of the soil involved is crucially important, and I suspect my generality could be contradicted by people living in an area where surface soil is darker than the soil just below.

Despite my limitations in being able to tell you much about it, the slight change in color is a very big factor in causing you to notice a track in the dirt, and you should think in terms of looking for these slight changes in hue.

A change in color is probably the most important factor when you are trying to track through green vegetation. Generally the track will appear lighter in color in short grass and darker in color (due to shadows) in thick, taller grass.

Also, I believe I can safely say that the underside on green leaves and blades of grass are lighter in color than their upward sides.

When plants grow naturally they do not turn their undersides upwards towards the sunlight. Therefore, when something lighter in color than the surrounding area catches your eye, and close inspection reveals it to be the underside of a leaf or blade of grass, it is pretty safe to assume that the vegetation has been moved by a force more powerful than it usually encounters in its natural environment.

Injuries to Vegetation

Bruising of vegetation is another important clue to look for. Greenery that is crushed between two hard surfaces will have a darkened, and possibly bleeding, wound. Since the shoe of the walker will provide one hard surface, you should scrutinize closely any greenery that is near other hard surfaces like rocks and sticks.

Greenery that is pressed against ground having a sandy surface will have grains of sand imbedded in its underside. The injury will be visible from above in the form of small dark bruises which give the greenery a speckled appearance. A close inspection (by turning the vegetation over to view its underside) will often reveal some grains of sand still imbedded in the leaf.

Bruised leaf with imbedded grains of sand.

Of course, a fresh, pliable, unwithered piece of greenery that is no longer attached to the plant is a very suspicious bit of sign and should be closely inspected to see if it was separated from its stalk by the sharp teeth of some small animal. If not, then it was probably knocked off by a rather large animal. This certainly makes it possible that it was caused by man.

Feeling for Sign

Areas with lush vegetation occur in parts of the country that have quite a bit of rainfall. This dampness can assist tracking as well as hamper it. Lush vegetation returns to its normal position with the passage of time and can thereby erase a trail. However, the indentation of the walker's heel remains impressed in the moist soil beneath the vegetation for a much longer period of time. If only one footprint can be found to give a good tracker a starting point, he can explore the ground with his fingertips and follow the trail by feeling these heel indentations. The utilization of the tracking stick greatly speeds this type of tracking operation.

I have found this feeling technique very useful on damp, leaf covered forest floors. While taking part in a training exercise in Great Smoky Mountain National Park, an area whose vegetation and wildlife were unfamiliar to me, I often resorted to the feeling technique to assure myself that what I was seeing was indeed the heel indentation of the person being followed.

Choosing Your Task

You have now been presented with 27 exercises dealing with sloping ground and vegetation. They do not have to be done in numerical order, you may do them in any order you wish. All that is required is that you do them all before progressing to Phase III exercises.

Tracking is great fun. I believe you will find that learning to track is great fun, and being able to select the problem that most appeals to you should add to the enjoyment.

Uphill tracking in lush forested area.

Author feeling for heel indentations on leafy forest floor.

7

Aging

Phase III Exercises

There are many forces that work to restore the land to its original form once it has been scarred by the movement of a human being. Knowing what some of these forces are, how they work, and how fast they work allows us to more accurately determine when the person we are tracking made his mark on the landscape.

The best way to tell the age of a track is by determining what it does to live vegetation. Grass, plants, bushes, and trees are all alive and like all living things, when injured, immediately start repairing themselves.

We are all very familiar with the healing process in human beings and can easily tell an old scar from a fresh bleeding wound. We can tell fairly accurately by the scabbing process if a cut has occurred within the last two hours, but the older the injury the more difficult it is to set time limits on when it might have occurred. We can usually tell if it falls within a time frame of more than twelve hours but less than three days old, probably more than three days but less than two weeks, etc.

A very similar healing process takes place with live vegetation and by studying the particular vegetation in your area you can easily learn to tell if an injury has occurred within a few hours, from several hours to within a day, etc.

The important thing to remember in deliberately injuring vegetation in order to study its healing process is to keep the damage inflicted as nearly as possible to that which normal walking would cause. In other words, do not grind your foot into the vegetation, simply walk over it. Do not uproot plants with your hand and lay them in the sun, as these will die and certainly wither far faster than plants that are kicked or squashed by normal walking and which still have all or part of their roots in the ground. These plants are still receiving sustenance from the earth and should live and return to a healthy state.

Avoid tearing leaves or branches completely loose from their bush. The natural act of walking will likely only tear them partially loose leaving the greater part of them still attached and being nourished.

Leaves which grow close to the ground are particularly more apt to be bruised than torn loose, so care must be used to injure them exactly as the natural walker would.

In testing how fast green grass will return to an upright position after being flattened by a human foot, you should realize that the taller the grass, the harder it is for it to stand back up. Short grass springs back much more quickly. Also the moisture of the ground plays an important part. Ground that is wet to the point of being muddy gives the roots of tall plants a very poor footing and causes the plants to take longer to return to an upright position.

Southern California has much dry, brown, dead appearing vegetation that when stepped on will remain flattened for weeks. People and animals walking across this dry grass leave a criss-crossing myriad of paths that seem impossible to differentiate from fresh trails. However, two things happen to the older trails that will cause them to have a slightly darker color than the fresh trail: First, moisture of dew and fog at night causes the once flattened stems to expand to their normal round shape which do not reflect sunlight as well as freshly flattened stems; and second, lying flat for a period of time allows dust to accumulate on the surface of the flattened area to a greater degree than is possible on stems standing upright. Therefore, the cleanness of freshly flattened stems allows them to reflect sunlight better and gives them their faintly lighter color.

As the night moisture of dew and fog act to re-inflate dry stems it has the same effect on flattened dirt surfaces. Not only does this slight blanket of moisture restore the ground to a uniform color, but it causes a graininess to gradually occur in flattened dirt which eventually erases the telltale sign of man.

A tracker must be very attuned to the forces at work in nature, and be very cognizant of the low spots where fog is apt to have occurred and the higher spots that it likely did not reach.

So too, he must be aware of the area's inhabitants. If your subject happens to walk a cattle trail at the wrong time of day (for you, the tracker) the movement of only three head of cattle can completely obliterate his trail in less than thirty minutes.

Smaller game, coyotes and rabbits, can obliterate a trail just as thoroughly, but it will take several hours. Field mice, snakes, birds, and ground crawling insects are abundant even in the most forbidding environments and their constant movement through an area soon erases all evidence

Footprint beginning to be erased by animal tracks.

of the track of man.

Creatures that occur very commonly in all environments are ants. Since ant hills are soft areas that are free of vegetation they appear to be excellent places to look for human footprints and if the person you are following stepped on the ant hill very recently, you will be able to see the track. However, ants like the shape of their ant hills or they would not build them that way. They are also tireless workers who begin immediately to restore their mound to its proper shape. The likelihood of a human track being recognizable in an ant hill twenty-four hours after the track was made is very unlikely.

To age tracks effectively, you must know a great deal about the number and type of creatures that inhabit the area where you do your tracking. Equally important is knowing whether they do their foraging in the early evening, at night, in early morning, or during the day.

This alone is the main reason I advocate teaching people to track in the area where they will be doing their tracking. Types of vegetation, weather conditions, the type of creatures inhabiting an area, and their habits are crucially important to the aging of tracks and it is therefore impossible for the greatest of trackers to be the unchallenged expert in every area. I have tracked or taught tracking from the Cascades to the Smoky Mountains, and have searched for lost persons from the High Sierras to the coasts and deserts of Baja California. Many times in these unfamiliar areas I have had to rely on experience and cunning to carry me over a vast expanse of personal ignorance concerning the aging factors at work in these strange surroundings.

Very early in our teaching program I began to realize that we should not be teaching people to track in an area which was strange to them. Therefore, I began to discourage groups from coming to El Cajon in order to learn our system. Far more people can be reached and far more good can be accomplished if you learn tracking and do your practicing in the area where you are going to be doing your searching.

Keeping in mind that the specifics of determining the age of tracks in your area must be learned in your area, I will offer a few generalities that should assist you in your study:

Determining Age of Broken Vegetation

A lot of sign noticed by a good tracker is not on the ground at all but appears from a few inches above the ground to the head height of the person being followed. Heavy brush is one of the very easiest terrain features for a tracker to follow someone through. Small broken branches from still living bushes and trees not only show from which side the pressure was applied which caused the break (which indicates direction of travel) but allow the tracker to examine

An obvious set of tracks beside a plain set of tracks. Both sets were made at exactly the same time. Most novices will generally identify obvious tracks as being fresher.

the color of the break in order to determine when it occurred. Breaks of this type get darker with the passage of time. The fresher the break, the lighter it will appear. The beauty of brush aging is that if you have doubts, all you have to do is break another small branch on the *same* bush and compare the color of the two.

This habit of frequently making comparisons is a good one for a tracker to develop very early. Expert trackers are continuously checking the way their own track is scuffing or marking the terrain in order to determine how much sign their quarry should be leaving.

Heat as an Aging Factor

Extreme heat increases the rate at which tracks appear to age. For tracks appearing in dirt, one of the most

Fresh break caused by leg being swung over fallen tree.

important things to look for is the change of color caused by the drier surface of the ground having been moved to expose a moister, lower surface. High temperatures work very swiftly to dry out this under surface and can restore the ground to a uniform color in as short a period of time as two or three hours.

As far as vegetation is concerned, heat is generally very beneficial to growing things so it does not alter the aging process in green vegetation greatly except in those cases where the bit of greenery has become completely separated from the plant and is in the process of withering. Extreme heat will greatly increase the withering rate, making a dislodged leaf which may only have been detached for a few hours look like it has been detached for more than a day.

Effects of Wind

Wind certainly erases sign quickly and can make very fresh tracks look quite old in just a few hours. A fact that beginning trackers often fail to consider is that nature is not the only producer of wind.

Tracks made alongside a busy roadway can seldom remain visible for twenty-four hours due to the many strong gusts of wind produced by passing vehicles. A road with heavy vehicular traffic, such as a freeway, can generate enough wind so the tracks within twenty or thirty feet of the roadway can be erased in less than six hours.

Effects of Rain

Rain obliterates tracks in dirt and sand completely unless it is only a light mist, and even light mists do a great deal to obscure sign. Light rains also encourage grass and weeds to stand back up more swiftly, thereby giving you less time to find grass trails. Dead brown grass which can only be tracked through because you are able to see a "shine", or the lighter color caused by light being reflected from its freshly flattened surfaces, is also taken away from you as a potential tracking area by the appearance of rain. Flattened stems are softened and return somewhat to their original round shape and the washing effect gives all the grass a uniform color.

One of the very few types of tracking area not affected by rain is brush and trees. The broken branches remain unchanged and the trail is just as obvious as if no rain had occurred.

If your quarry is still moving during the rain he will transfer mud from his shoes to the tops of blades of grass, rocks, and even paved roads or sidewalks. However, if the rain is continuing, it will immediately start washing this mud away.

If your quarry is moving after a rainstorm you have

all the cards in your favor if you can keep the area free of non-tracking track-makers. All sign, human and animal, will be washed away by the rain and you have a fresh slate on which to start. The movement of animals cannot be controlled but their sign is very easy to distinguish from man's sign anyway. What you must strive to do is keep other people out of the area where you wish to do your tracking.

Ideal tracking conditions occur after rainstorm.

Problems With Snow

Snow is a weather condition like wind and rain that immediately starts obscuring your existing trail. However, it provides you with a clean slate that could make your tracking problem ridiculously simple.

Snow can be the somewhat damp variety that in the northern states commonly falls in the spring and fall seasons; it can be the dry, powdery variety that is referred to as winter snow; or it can be the tiny, ice-ball variety that midwesterners call sleet. It can occur with high winds or with very little wind, at times when the temperature stabilizes below 32 degrees Fahrenheit, or at times when the temperature fluctuates above and below freezing.

These three factors; type of snow, amount of wind, and fluctuation of temperature, work in various combinations to make snow tracking range from easy to impossible.

Given stable, below freezing temperatures and no strong wind, the tracker will prefer to track in damp snow because he will be apt to get a "make" on the tracks, (make is tracking jargon for positive description of the track). This means that he will be more able to follow a particular set of tracks through a maze of other similar trails with little chance of confusion.

Powdery snow does not compact as well as damp snow, thereby lessening the chances for the shoe's distinctive markings to have been preserved. Also, the snow around the edge of the track falls in almost immediately when it is powdery, covering any distinctive markings that might have been visible. A technique that will sometimes allow you to re-expose a make in powdery snow is to clear away the high sides of snow surrounding the track with your hands so that it cannot continue falling in, then blow on the track to scatter the loose flakes that are lying on top of the compacted surface.

When you add high wind and new snow to either a damp snow trail or a powdery snow trail, you are probably going to have your trails obliterated. However, if you only add high wind, your damp snow trail will very likely remain while your powdery snow trail vanishes.

The addition of new snow and moderate wind could result in your primary evidence being filled in, but your secondary evidence remaining, thereby still allowing you something to follow. When a person walks in snow they make a deep impression but their foot also picks up small blobs of snow that drop off just forward of the footprint. Sometimes, with the right conditions, the indentation will be completely obscured by new snow while the tiny mounds of dropped snow will be built upon, staying visible much beyond the point where the indentations will have disappeared.

The most confusing problems with snow tracking occur when temperature fluctuations allow thawing and re-freezing to occur. Thawing, of course, greatly distorts the track making it appear very old, or actually making it become unrecognizable as a human track. The problem that this presents is more severe in signcutting than in tracking.

In tracking, the continuity of your trail will carry you through areas where the track has remained shaded and thus relatively unchanged, then into areas where long exposure to direct sunlight will have altered the track greatly. Since this distorted mess occurs exactly where your tracks should be appearing (at regular step intervals) a good tracker will continue to follow it without experiencing great difficulty.

You must also consider that in winter in the United States the sun's light approaches earth from low in the southern hemisphere, therefore, tracks on the south slope of a hill will get more direct sunlight than tracks on flat ground and considerably more than tracks on the north slope of even slight rises in elevation. This means that the same trail will look older in those areas where it has received the most direct sunlight.

You can surely understand the problems this creates in trying to cut sign. The inexperienced tracker will always find "fresh" tracks in shaded areas and always find "old" tracks in sun drenched areas.

Some degree of accurate aging can be done in damp snow due to the fact that freshly compacted areas will be snow-white and as they age will become ice-gray. The point of greatest foot pressure should be finger tested to determine if it has started to re-freeze. If your finger breaks through this point easily, you have a fresh track. As the track ages it will slowly freeze outward from the points of greatest pressure to the points of least pressure turning from white to ice-gray as it does.

Again, as I have said many times, make a comparison with your own track, or make the comparison with the track of another person who was in the area at a time known to you.

Sleet for the most part poses impossible tracking conditions. An exception occurs, however, if we have a thin layer of icy sleet covering several inches of softer snow. In this instance the weight of the walker may occasionally break through the icy crust leaving you a few intermittent tracks. This condition forces even good trackers to resort to the distasteful practice of jump-tracking.

The intelligent determination of the age of a track is something that will take years to accomplish. You must look at a trail that is several hours old, then again after a day, then again after several days. You must do this in all seasons and you must do it under all wind and moisture conditions. There is no way to shortcut.

Like aging itself, it is just something that is going to take time.

PHASE III: FIELD EXERCISES

In all the preceding field exercises you have been working with tracks that had been made almost immediately before you went out and attempted to follow them. They were just about as fresh as possible, only being a few hours old by the time you were finishing the problem.

If you have diligently performed all the previous exercises you should have by now completed about 100 hours of tracking. It is questionable if that is enough practice time to prepare the average student tracker properly for the next phase of training which deals with problems of time, aging, and weather. If you find you are not ready for this phase all you have to do is get in more practice on Phase II problems. If you feel that you are an accomplished tracker you may pick terrain that is more difficult than what I have outlined below. However, I believe most of you will be severely challenged by the four Phase III exercises that follow:

1. Side hilling (keeping the same elevation while walking on sloping ground) in an area with vegetation and ground cover, tracks heading toward the early morning sun, tracks laid down before sunset the evening before you intend to follow them (12 to 15 hours previously).
2. Same as above except tracks laid out just after dawn of the previous day (24 hours earlier).
3. Same as above except tracks laid out 36 to 39 hours earlier (before sunset a day and a half before).
4. Same as above except tracks laid out 48 hours earlier (just after dawn two days before).

You should work on each of these problems at least three times for a period of about three hours each before moving on to the next problem. It is highly advantageous to have at least three different training sites that meet the desired criteria for this set of problems so that you work "hillside A" the first time you do problem number 1, "hillside B" the second time you do problem number 1, and "hillside C" the third time you do problem number 1.

This accomplishes two important purposes; it adds greater variety to the vegetation and ground cover you must learn to interpret, and it allows an adequate healing time for the site to restore itself to its natural condition so as to be ready for the next exercise.

Three sets of tracks in soft sand. Tracks on left were made Sunday, tracks in middle on Monday, and tracks at right on Wednesday of the same week.

Three lugged-sole tracks in soft sand. Tracks were made at exactly same time as flat-sole tracks on opposite page. Sunday track in middle, Monday track on right, Wednesday track on left.

8

Signcutting

Signcutting is the act of looking for sign in order to find a starting point to begin tracking. At this point we are going to assume that you have completed all the practice exercises and are ready to test your skill by discovering an existing trail and attempting to follow it.

When to Look

Under ideal circumstances you should begin your search for sign at sunrise and continue it while the sun is rising in the first one-third of its daily arc. Under non-emergency situations you should then suspend your search for sign during the mid-part of the day and not resume it again until the sun enters the last one-third of its downward arc prior to sunset.

Neither in law enforcement nor in searching for lost persons do you have this luxury of time that allows for such

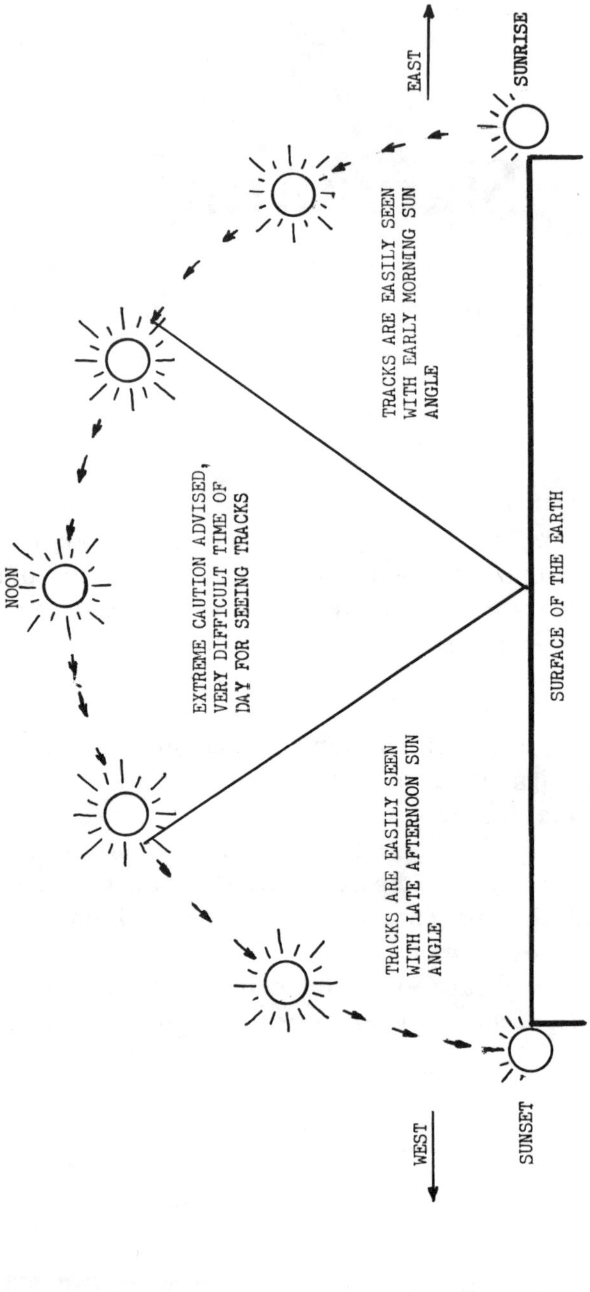
EAST
SUNRISE
TRACKS ARE EASILY SEEN
WITH EARLY MORNING SUN
ANGLE
EXTREME CAUTION ADVISED,
VERY DIFFICULT TIME OF
DAY FOR SEEING TRACKS
NOON
SURFACE OF THE EARTH
TRACKS ARE EASILY SEEN
WITH LATE AFTERNOON SUN
ANGLE
SUNSET
WEST

a long respite at mid-day. Therefore, it is only to emphasize the point that mid-day is a terrible time to look for sign that I facetiously suggest that you not even try. It is just that I want to strongly discourage anyone from driving a vehicle on a dirt road during this period of the day if there is a possibility that the sign of the person you seek might appear there. Walk the road (keeping to the edge away from the sun but looking towards the sun) send a *single* horseman along the road (keeping to the edge away from the sun but looking towards the sun) or send a *single,* slow riding motorcyclist along the road (keeping to the edge away from the sun but looking towards the sun) but strongly avoid sending a four wheeled vehicle onto that road.

Vehicle tire tracks are the number two destroyers of sign, ranking very close to the number one destroyer which is the foot traffic of non-trackers. Therefore, the movement of vehicles along dirt roads must be controlled very carefully. Wind, weather, and the movement of small creatures all destroy sign but not nearly so quickly or so devastatingly as vehicles and people.

As you will be forced to look for sign at mid-day you will also, at times, be forced to signcut at night. In the Border Patrol we expend more man hours searching for people at night than we do in daylight. I in no way intend to infer that we spend more time tracking at night, this is not so. We do the preponderance of our tracking during daylight. However, we probably do a little more signcutting at night than during the day so we have had to make some adjustments in order to enhance our chances of seeing sign under such poor light conditions.

We have equipped our vehicles with tractor lights that are hinged to the bumper and can be swiveled downward to within a few inches of the ground in order to cast a light which produces the shadows that reveal slight indentations in the ground. Of course the vehicle headlights must be turned off when utilizing these signcut lights because their high angle light eliminates the desired shadows. These signcutting lights are far better than regular vehicle lights, but they are not nearly as effective as natural light. Their weakness stems from the fact that you are forced to be behind the light source instead of looking towards it. However, if they are used cautiously they can be effective.

Tracker adjusting signcutting lights.

My strong personal preference for night signcutting is a slow moving motorcycle with low slung lights. You are closer to the sign, your field of vision is better, and if the sign is missed by the signcutter the vehicle tires will likely not obliterate all of the evidence, as is often the case with four-wheel vehicles. If you are cutting for sign on foot you should use a good wide-beam flashlight held close to the ground and as far to the side as possible so as to cast a light directly across your path.

Night signcutting.

Signcutting from a vehicle. Cameraman shooting with sun to his back.

With correct sun angle the camera can pick up what the signcutter is seeing.

How to Look

In cutting for sign you should strive to intersect the path of the person you are seeking at a 90 degree angle so as not to confuse the trail with your own sign. If you are on foot it is very prudent to drag your signcutting stick. This tells all subsequent searchers who see your tracks that you are a searcher and not the subject being sought.

The most important thing to be said about how to look for sign should by now have been hammered into you by the Phase I exercises:

> Look for the track to appear directly between you and the light source. Regardless of which direction you are going, have your head turned and your eyes positioned so as to scrutinize the patch of ground that is directly between you and the sun.

What to Look For

Specifically you look for everything mentioned in Chapter 5, but generally you look for:

1. Flattening
2. Regularity
3. Color Change
4. Disturbance

Where to Look

First look in those areas where the person being sought spent enough time that he should have left his footprint; second, look along a route that the person you are seeking might have been induced to walk; and third, look in those areas where a track would be very easy to see.

The first concern is always to establish which track among hundreds is the one you should be following. To do this, interrogate whatever people might be present in order to determine a location where the subject walked or stood that might have been off the beaten path. If several of his

Flattening and regularity.

companions were with him at this site you should look at the soles of their shoes in order to know what tracks can be eliminated. If your search originates at an abandoned car, carefully inspect the ground beneath all doors.

Dirt Roads

The best places to look for tracks are the dusty, dirt roads in the area where the person you are seeking was last seen. However, do not place great faith in these roads unless you know with some certainty that they have been free of vehicle traffic since the time your subject would have walked there.

Sandy Streambeds

Sandy streambeds are very good places to look for tracks, but you must be cautious of two very common signcutting errors: The first is expecting the tracks you are

Signcutter cutting dirt road.

looking for to appear as fresh as your own. Realize that soft sand will not keep a track sharply defined very long. So if you are looking for the track of a person who might have been walking in that soft sand two days earlier be very certain that you know what a two day old track looks like in soft sand. Gravity does not tolerate sharply defined edges in soft sand; it works very swiftly to smooth and level. A track two hours old looks nothing at all like a track eight hours old. Be sure you know the difference! The second common error is in failing to realize that sandy stream beds frequently have patches of a very firm clay-like silt which is often so hard that it can only be scuffed and will not reveal a full footprint at all. From several feet away these patches look like smooth sand. You must walk on them yourself to check how your own track affects them.

Also, be aware of the fact that a smooth, flat track will be visible in sand much longer than an uneven, high

Freshness of track is evident by sharply defined instep brand (flat area) but sand has fallen in so fast in lugged sole area that track appears to be old.

traction track such as a Vibram sole. As was mentioned, wind and gravity work to level a track, but a smooth, flat track is level to begin with, so change occurs very slowly. Vibram soles and hiking boots create many high and low spots that can show the effects of nature's leveling action very quickly.

Animal Trails

There are practical reasons and probably strong psychological pressures that induce people to follow well defined game or domestic animal trails. When such trails are visible, they are obvious pathways, and people have a strong tendency to use them. Aside from possible psychological reasons, they are free of heavy brush and are therefore easier to walk. They also will eventually lead to water.

I can vividly remember one particular search which took me into one of Mexico's most dangerous deserts east of the Sierra De Juarez Mountains in rugged Baja California. In following the tracks of the lost person we were seeking, my partner and I climbed out of a deep wash and onto a high flat mesa that was a veritable obstacle course of every kind of cactus that I could imagine. It would have been a hopeless maze if not for the animal paths that led you through it.

After having followed our victim through several miles of trackless desert we were amazed by the steadily increasing number of animal tracks we were encountering. In a distance of less than a mile it became such a freeway of animal tracks that our victim's track was completely erased and we were forced to assume that he was still on the trail. We continued following the animal sign to one of the most unexpected, coolly shaded watering holes you would ever hope to find.

Our victim's track was not there and when we finally found it again it was in such a location as to indicate he had probably turned against the animal traffic and walked away from the watering hole.

Another very good reason, I thought to myself, why people who go into the wilderness should know a great deal more than they do about tracking. Had our subject been conscious of these animal tracks he probably would have followed them to water and been saved.

Natural "Track Traps"

Other naturally occurring places that are outstanding to check for sign because of the ease with which tracks can be seen are: mud flats, salt flats, fields of high grass, river banks, bushy thickets, and steep embankments.

Any steep slope is good because (as was mentioned in Chapter 6) in order to walk up, down, or across a steeply inclined plane a person must plant his foot firmly, sometimes creating a little terrace but always inflicting extensive damage to the surface of the ground in order to maintain his footing.

Man-made "Track Traps"

Man-made "track traps" can be equally as helpful when looking for sign. In addition to dirt roads, we have bulldozed firebreaks, plowed fields, cleared construction sites, the berms of roadways, and the terrain along fencelines.

Getting over, under, or through a barbed wire fence requires extra foot placements and a degree of pivoting and scuffing which should leave a noticeable amount of sign.

Route Influencers

Fences also come under the category of things that will induce a person to walk a certain route. A person lost in fog will be particularly thankful for something like a fence that will lead him in a straight line.

Pipelines, electric lines, and telephone lines are all apt to influence a person's route since surely they will

Signcutter cutting steep embankment.

lead him to civilization.

City lights, airport beacons, distant water towers, or simply prominent mountain peaks offer goals or points of reference to a lost person and are frequently selected as guide points. Once it appears evident that the person whose tracks you are following is being influenced by such a landmark you can intelligently select a point, a number of miles ahead, where it would be prudent to cut sign for his tracks.

Cutting Ahead

The intention of signcutting is to find a track in order to begin a tracking operation. Once the tracking operation is under way the same process is re-employed to speedily advance the tracks great distances, but this time, in tracking parlance, it is called "cutting ahead".

Cutting ahead begins with the tracking team "lining

out" the probable path of the person being followed. This simply means that they follow his tracks until they feel they know what his intended route is going to be and then they radio this information to the teams that are cutting ahead. If they feel his intention is to simply walk east they will so advise; if it appears he is walking a particular canyon, guiding on a specific landmark, or following a power line, this is the information that they relay ahead.

With this information the forward teams may proceed by vehicle directly to points along this probable route, judiciously select terrain that will readily reveal sign, and carefully make a cut.

The team making the cut a mile or so ahead may not even get in position before another team, three miles ahead, reports finding the tracks.

At this juncture the forward team abandons its vehicle to become the tracking team while the rear teams leapfrog ahead to make another cut.

Once the tracks do not appear where it is felt they should, the second most likely route is cut, then the third most likely, and this continues until the absolutely unlikely routes are checked. This usually results in a complete perimeter being cut around the steady advance of the tracking team.

As the areas of highest probability are eliminated, the forward cutting teams draw closer back to the tracking team in ever tightening circles. It is quite common that when the subject is actually found the forward teams have closed the circle so tightly that they are usually on foot and within shouting distance of the successful tracking team.

9

Track Identification and Description

The thing that makes the greatest impression on beginning trackers is the sudden vast number of tracks that they are able to see when before, they had noticed nothing. In this same vein, they are soon overwhelmed by the boundless variety of markings that appear on the bottoms of shoes.

My own awareness of the magnitude of this problem dates back to 1967, when, while working with the F.B.I. on a kidnap-murder case whose victims were Federal law enforcement officers, we were supplied with photographs of heel patterns that appeared only on American-made shoes. There were over 2,000 photographs dealing with this single category. When you add sole patterns, tennis shoe patterns, hiking boot patterns, and foreign-made shoe patterns you have a subject of encyclopedic proportions.

Fortunately it is not at all necessary to be able to instantly recognize and catalogue a vast number of shoe prints. All a tracker must be able to do is:

1. Differentiate the track from similar tracks
2. Remember what it looks like
3. Communicate the track description to co-workers

Both of these first two requirements can be easily complied with in one operation by studying the track carefully, measuring it, and drawing a picture of it. Photography is excellent and I advocate it, but nothing acquaints you as thoroughly with a track as does painstakingly measuring it and penciling in all of its unique markings.

As mentioned in Chapter 3, you should measure the overall length of the foot, and the sole at its widest point. If it has a heel, you should measure its length and width, note the number of nail holes, note any stitching, note the number and sizes of lines and markings, and you should record all of this information on your drawing.

Lastly you should look for cuts, worn spots, or other idiosyncracies that make the track unique.

If you have an instant-print camera available you should certainly photograph the print and distribute the photos to everyone involved in the search, but do not use this as a substitute for your drawing.

One photograph will never show all the peculiarities of a single track. Information regarding the track's complete appearance normally has to be compiled bit by bit as a result of seeing many prints and gathering separate information from each. For this reason your ultimate drawing will provide you with more information than a single photograph.

Your photograph is a great way to communicate your track description to co-workers. However, so is a photocopy of your drawing. If you have these devices available to you, by all means use them. However, most of the time you will find yourself being forced to paint a word picture over the radio.

In describing tracks by radio you should keep it

simple; begin with the general description, and proceed to the specific details.

The officers with whom I work, the professionals who handle numerous track descriptions every working day, rarely bombard me with extraneous or unnecessary information.

Trackers describe tracks that have no separate, elevated heel as "flats". Tracks that do have a separate or elevated heel are called "heels". These terms may have to be explained when communicating with a non-tracker, but among trackers they do fine.

Normally, the first thing one of my fellow officers will tell me is that they have either a "flat" or a "heel". There is no need to amplify on this. Tracks either have a heel or they do not, and I immediately start getting a general mental picture of what I am to look for.

Since the preponderance of the tracks we are looking for belong to adult males we do not bother to point this out each time. It is assumed that it is an adult male unless you are told otherwise.

If it develops that the particular track description being radioed to me is one of those exceptions when it belongs to a woman or a child, then this fact is the first one reported to me, and whether it is a heel or flat comes next.

Flats

Basically there are three main categories of flats; they are: (1) Tennies, (2) Sandals, and (3) Work Boots.

Trackers refer to all manner of flat-soled athletic shoes as "tennies", meaning "tennis shoes". With this one word they tell you that it is a flat; that it falls into the category of athletic shoes; and that it is going to have a variety of descriptive markings on its sole. Athletic shoes are designed for high traction which means they must have raised areas on their soles which will leave noticeable marks on the ground. They are also designed to protect the foot during periods when it is subjected to straining and twisting motions. For this reason it is well contoured to the shape of the foot and therefore is easily recognized as a tennie by its shape even after all distinctive markings are worn off.

Sandals on the other hand, are very often smooth. If they have a sole pattern at all it is most likely to be shallow and uncomplicated.

Sandals are seldom narrowest at the instep, as are tennies. They usually maintain the heel width through the instep area then spread at the ball of the foot and sweep towards a point in the area of the big toe.

The shape of work boots is typified by heel and instep areas that are fairly wide in relation to the overall length of the track. They generally have a rounded toe. The sole markings are generally less ornate than tennies and are coarser in appearance.

In each of these categories of flat-soled shoes there are a number of commonly appearing tracks that allow the person describing the track to communicate a great deal with very few words. "Herringbone tennie" instantly tells you to look for a flat-soled tennis shoe with fine wavy lines. "Fine lined tennie" tells you it is a flat tennis shoe with straight, fine lines.

Two commonly appearing tennis shoes worn by aliens who attempt to enter the United States in our area lend themselves beautifully to apt description; they are the "ball bearing tennie" and the "torch tennie".

The "ball bearing tennie" has clusters of small circles throughout the track and the "torch tennie" is a Canada brand which shows a hand holding a torch in the heel area.

The most commonly appearing sandal is simply called a "shower shoe". This tells you it is flat, shaped like a sandal, and may reveal the three circles that anchor the thongs to the sole of the shoe.

My tracking colleagues describe another common

sandal, which has a tire tread sole, by its single-word Mexican name which is "Huarache".

The most common flat work boot is sold by Sears, Red Wing, and a number of other companies. It is recognized by thick lines, referred to as "bars", which are interrupted as they cross the shoe. This interruption is either a blank area or an offset bar. Hence, they are commonly referred to as "broken bars".

A sub-category of flats that is extremely popular at this time is the Earth Shoe. These are very distinctively shaped having an appearance much like a blunted sandal, but are most noticeable because the lines in the heel area are not parallel with those in the sole.

Heels

Shoes with heels are usually lumped into only two main categories: Street shoes/dress shoes and hiking boots.

Street shoes are described first by the markings in the heel unless there is a more prominent recognition point such as the shape of the heel, the shape of the sole, etc.

A person who is good at describing tracks will tell you very quickly about the most noticeable aspect of the track. Usually with street shoes, the most noticeable aspects are the pattern in the heel, the number or spacing of nail holes, or the shape of the leading edge of the heel. However, if the heel is relatively plain, the most unique characteristic of the track might be a sharply pointed toe, stitching around the sole, or possibly a hole in the sole.

A few years ago we experienced one of those rare times when shoe designers succeeded in establishing an outlandish fad fashion in men's footwear. The "mod shoes" were characterized by high heels which had concave leading edges and toes which were rounded. These still appear and are called "mod shoes" or "fruit boots". I suspect that all future fad extremes in men's footwear will be referred to by one or both of these terms.

The most common hiking boot is the lugged Vibram sole. It has been so successful that all types of cheap imitations have sprung up. The common discount house varieties are generally described as "waffle stompers" while the more expensive imitations are referred to by my tracking colleagues as "Mexican vibrams".

In the hiking boot category are two types issued to military personnel, the "Vietnam boot" which might otherwise have fallen into the category described as "Mexican vibrams" and the "running W".

The "running W" is comprised of heavy wavy lines and is distinctly different from the "herringbone tennie" because of the thickness of the lines and the fact that the track has a raised heel.

Descriptive Terms

In describing either flats or heels there are unlimited descriptive words and phrases that convey vivid mental pictures. Circle, square, diamond, and football describe the more obvious shapes while line, bar (thick line), broken bar, wavy line, herringbone, and running W aptly describe lines.

For genuine imaginativeness and real word pictures though, I have to admire terms such as: "screen wire effect", "waffle pattern", "quilted pattern", "boiler plate design", "basket weave", "heel within a heel", "heavy chevrons", "rope sole", "honeycomb", "diamond with chevrons", "sunburst", "spider web", and "snake skin" just to name a few.

The Acme Boot Company puts a brand on the heels of its boots that looks like a cigar band would if you unstuck it and laid it out flat, hence "cigar band" tells a person who is unfamiliar with the track exactly what he should be looking for.

The most important advice I can give you is keep your descriptions in terms that everyone can relate to. You reach far more people by saying that you have "a football" in the instep than you do by saying you have "an ellipse".

Measurements

Tracks should be accurately measured and this information should be carefully recorded. Geometrical shapes should be counted, as should lines, and the distance separating them should be measured. However, this precision is only necessary to keep you from becoming confused when you encounter an almost identical track.

The only measurement information with which you should tie up the air is overall length of the shoe, and if it has a heel, the length and width of the heel. The rest of your measurements should only be transmitted after someone has found a track which they feel matches your initial description.

Remember it is far more communicative to tell a person that the track has a circle in the heel "about the size of a quarter" than to tell them it contains a circle "5/16 inches in diameter".

It is far more useful to tell them that lines are "about the thickness of a match stick" than to tell them you have lines "1/8 inch thick"

A problem with measurements of which you should be cognizant is the fact that measurements will vary depending on the consistency of the soil in the area where the measurements take place. In soft, dry sand you will always get only an approximate measurement because the edges of the track fall in the instant the foot is removed from the sand. In mud an imperceptible sliding effect can lengthen your measurement as much as half an inch.

If the soil conditions fall in some of these unreliable categories be sure to communicate what type of soil you took your measurement in when you relay the rest of your track description.

Shoe Sizes

All long-time trackers bristle when they hear a track described as "a size 9 track" or "about a size 10". There is no way in the world to accurately tell the shoe size of a track by its overall measurement. People who attempt to describe tracks in this manner probably know what their own shoe size is and since the track they are describing is the same size, they are simply trying to impress someone with their saltiness.

Shoe manufacturers vary widely in sizing shoes. It is an unusual person who owns well fitting shoes made by six different manufacturers and has all of them sized the same. As far as overall measurement is concerned, I find a difference of more than one-half inch between shortest and longest when I measure six of my own shoes.

Several years ago I was called upon to help find a mentally retarded boy who had wandered away from a ranch where forty mentally retarded boys were living. We discovered immediately that the same tennis shoe pattern featuring a diamond design was everywhere.

A quick check with the counselors revealed that tennis shoes were bought by the gross from a single manufacturer and issued to the boys as their old ones wore out. It was immediately apparent that every boy on the ranch had the same track except for differences in size. A check of camp records quickly supplied us with our victim's shoe size so we started measuring new shoes in order to translate this shoe size information into a length that would be meaningful to us. To our amazement we discovered more than a quarter of an inch difference in the overall measurement between two shoes of the same size which were made by the same manufacturer. Therefore, we had no way of differentiating the track of the boy we were seeking from at least a dozen other boys' tracks.

Even if this discrepancy between shoe size and measurement did not exist, I find the use of shoe size to be a poor communicator. None of the trackers with whom I associate carry around a "shoe sizer" which they can flop down beside a track in order to see if it is the "size 9" that has been called to them. However, many of us carry small measuring tapes and can quickly ascertain if a suspect track has the same measurements as the one we are seeking.

Walking Traits

Before we leave the subject of track identification some mention should be made of walking traits.

The common terms trackers use to describe the way a walker turns his feet are "pigeon toed" for those whose toes turn inward and "splay footed" for those whose toes point outward.

When one foot consistently takes a shorter stride than the other you are probably following someone with a limp.

A dragging mark to the rear of heel placements indicates that the walker is tiring. Complete shuffling of both feet indicates a walker who is near exhaustion.

Box-like foot placements (short strides coupled with a wide side-to-side spread) indicates the careful and labored walk of the elderly.

An extended stride indicates trotting or jogging while an extremely extended stride showing the weight mostly on the balls of the feet indicates a full-speed sprint.

In addition to looking for these very helpful characteristics you should also be on the lookout for habits or personal preferences such as a tendency to walk to one side of a road; a proclivity to walk around a small bush instead of through it; an aversion for dense woods; a preference for walking a drainage instead of a ridgeback; etc. All of these elements are a vital part of track description. They should be noted and communicated.

Tracking is a highly visual art. Communication is largely verbal, and when radio communication is involved it is entirely verbal. It is in order to drive home that point that I chose to have this chapter be the only one in the entire book that does not contain a single photograph or illustration.

10
Following a Non-Visible Trail

In search and rescue work far more than in my Border Patrol experience, I have frequently been called upon to follow non-visible trails.

A non-visible trail is the extension of an actual trail that has been erased by nature or man.

Many searches occur when children wander away from campgrounds. Often these are areas which have much confusing foot traffic to begin with. Add to this the tracks of the immediate family who does some searching before they notify authorities. Sprinkle in the tracks of the initially contacted authorities who do preliminary checking before they call for an all-out search. Douse thoroughly with the tracks of well meaning, fellow campers who have heard a child is lost, and finish off with the caravans of searchers' vehicles drawn into the area to launch the search, and you present the tracker with a problem of almost insurmountable magnitude.

If the tracker can identify which track belongs to the lost person, and if he can find that track beyond the perimeter that has been trampled by non-tracking searchers, he will probably follow that track into an area where it will disappear under searcher's tracks, vehicle tracks, or animal tracks. Wind or weather may also eliminate the trail or it might go onto a paved road, onto solid rock, or onto a grassy area that has had time to return to its original appearance.

In these instances the tracker is left with the choice of cutting ahead or following a non-visible trail.

Following a non-visible trail is advisable when the tracker can be sure that the person being followed could not leave the trail, to either side, without the tracker being able to see evidence of it. Following the non-visible trail is done when, to continue on, would not add confusing sign to the terrain (across rocks, pavement, etc.) and it is done when the

Two trackers continuing trail on paved road while warily cutting each side for departure point.

tracker is absolutely certain that no amount of tracking skill beyond his own would enable someone else to follow the trail.

If doubt exists as to any of these three conditions, then cutting-ahead (as outlined in Chapter 8) should be done.

Following a non-visible trail is done by process of elimination and logical thinking.

> (Example—The person we are following started this way but physical evidence of his passage does not now exist. If he exited to the right he would have had to climb that steep bank which would show us much visible evidence. If he exited to the left he would have had to push his way through dense brush which would show us many freshly broken branches, hence he is still on this trail.)

Several years ago in searching for some children who had failed to return from a hike in the area of Palomar Mountain, we followed their tracks through the night and arrived near dawn at a footpath which descended into the treacherous Pauma Creek area. The tracks of the lost children went right onto a dusty path where they were completely covered by the tracks of six adult searchers who late the night before had gone to the canyon bottom and returned by this single trail.

For more than two miles my partner and I followed this non-visible trail. My partner made sure that the children had not gone to the left and I made sure they had not gone to the right. Eventually we reached a point where the searchers had turned back and where we were immediately rewarded for our perseverance by the sight of our subjects' tracks continuing down the drainage. Three hours later we found them, but the area was so rugged that it was almost six hours more before we climbed to a point where we could be met and evacuated.

Similar procedures are followed when a person gets into an area that must be traversed by stepping from rock to

Tracker following subjects indicated route across solid rock.

Tracker carefully checking cutable area at end of long rock.

Tracker following trail down rocky stream bed while warily checking easy exit routes to his right.

rock. If this was the direction the subject was proceeding, then this is the direction in which he will likely continue. Follow the easiest, most natural walking route. What is easiest for you would have been easiest for him. Eventually you will reach a point where you cannot easily step to the next boulder but must step on some greenery, some brush, a log, or the ground itself. Stop and look very carefully where you intend to place your foot and in all probability you will find the subtle evidence which indicates that this is exactly where your quarry placed his foot too.

In doing this kind of tracking the thing you must be diligent about is constantly checking both sides of the trail for evidence of a sudden turn. If you have the ideal situation of a three man tracking team this will provide a full time job for each of you.

However, if you are working alone you cannot afford to spend all your time being certain that no sudden turns were made. The high probability is that the person continued on his chosen course. People do not make radical changes in direction without good reason. If you see no reason for a dramatic change of direction then it is unlikely the person you are following saw one.

However, keep in mind that conditions which existed when your quarry was there may have changed by the time you arrive at that point. A fleeing lawbreaker may have skirted around some campers or a parked car and when you arrive these people may have departed.

A lost child may have altered course upon seeing a snake or a fearsome animal which is nowhere in sight when you arrive. So don't jump to conclusions about a person's mental condition just because you cannot explain their actions.

Try to rely on solid evidence, tracks and sign, but when these are taken from you, stick with the high probabilities and common sense.

Scuffed lichen on solid rock.

In the area of jumping to conclusions I have had inexperienced trackers tell me that the trail they had cut indicated that the person we were seeking was in abject panic. They supported their contention by pointing out that the subject was crashing pell-mell through heavy brush when there were unobstructed routes only a few feet to the side. Upon examining the breaks in the brush it was evident that they were probably caused between sunset and dawn. Remembering that there had been absolutely no moon during the night, led to the more accurate conclusion that the sharpest eyed person in the world could not have seen well enough to have found his way to those unobstructed paths.

As a general rule, the fact that a person crashes through brush, bumps into objects, allows their leg to contact a cactus, or trips over an exposed root strongly indicates that they are walking at night rather than that they have suddenly become addle brained.

A person shows panic or deteriorated mental condition most strongly by abandoning articles necessary for his survival or comfort. If you find no such articles along a person's trail you do not need to get alarmed about his possible mental condition.

NON-VISIBLE SIGN

Things that you hear and things that you smell come under the category of non-visible sign. We have already mentioned a few of the sounds to be conscious of, such as the barking of dogs or the sudden squawking of birds. You surely know without being told that there might be much significance in the discharge of a firearm or the sound of a whistle, but the probability is likely quite high that you do not pay a great deal of attention to unusual smells.

A number of trees give off stronger scents when their limbs are broken or when their bark is knocked off, and a great number of plants give off far stronger scents upon being trampled than they would under ordinary circumstances.

In order for this information to be useful to you, you must develop the habit of being aware of unusually strong scents in your area and being curious enough to determine their origin.

Like learning to determine the healing time of vegetation in your area, you must injure, examine, and compare in order to acquire the knowledge that will allow you to reach intelligent conclusions concerning the reasons for the sudden manifestation of a strong scent along the trail you are following.

I can remember any number of times when scent has helped me find aliens who in the course of trying to enter the country illegally had walked for several days through the mountains east of San Diego, building campfires as they went. As I followed their tracks towards their final hiding place, the scent from those campfires was so strong on their clothing that I found them easily despite the fact that they were extremely well concealed. Indeed, if it were not for the roughness of the terrain, I probably could have walked the final fifteen or twenty feet blindfolded.

Smells, sounds, feelings, and sights are what life is all about. Learning to track will make you far more aware of these things than you have ever been. Do not make the mistake of restricting this expansion of your awareness only to things that you can see. Strive to expand this awareness in all of your senses and you will find that learning to track will have enriched your life.

PHOTO CREDITS

Page no.	4	Joe Cordova
Page no.	9	Joe O'Dell
Page no.	13	Joe O'Dell
Page no.	15	Photo courtesy of Daily News, Indio California
Page no.	20	Photo courtesy of San Diego Mountain Rescue Team
Page no.	23	Joe O'Dell
Page no.	29	Joe O'Dell
Page no.	30	Joe O'Dell
Page no.	38	Joe O'Dell
Page no.	40	Mackintosh Studios
Page no.	44	Mackintosh Studios
Page no.	45	Mackintosh Studios
Page no.	51	Mackintosh Studios
Page no.	57	Top Joe O'Dell
Page no.	57	Bottom Joe Cordova
Page no.	58	Top Todd Purcell
Page no.	58	Bottom Joe O'Dell
Page no.	60	Joe O'Dell
Page no.	61	Joe O'Dell
Page no.	62	Mackintosh Studios
Page no.	63	Top Joe Cordova
Page no.	63	Bottom Joe Cordova
Page no.	67	Joe Cordova
Page no.	71	Joe O'Dell
Page no.	74	Joe Cordova
Page no.	76	Joe Cordova
Page no.	80	Joe O'Dell
Page no.	82	Joe O'Dell
Page no.	87	Mackintosh Studios
Page no.	88	Mackintosh Studios
Page no.	90	Joe Cordova
Page no.	91	Joe Cordova
Page no.	93	Joe Cordova
Page no.	95	Joe O'Dell
Page no.	96	Joe O'Dell

PHOTO CREDITS

Page no.	100	Joe Cordova
Page no.	102	Joe Cordova
Page no.	103	Joe O'Dell
Page no.	105	Joe Cordova
Page no.	110	Joe O'Dell
Page no.	111	Joe O'Dell
Page no.	116	Joe O'Dell
Page no.	117	Photo by Joe O'Dell—Mackintosh Studios
Page no.	118	Joe O'Dell
Page no.	119	Joe O'Dell
Page no.	121	Joe O'Dell
Page no.	122	Joe O'Dell
Page no.	123	Joe O'Dell
Page no.	126	Joe O'Dell
Page no.	140	Mackintosh Studios
Page no.	142	Mackintosh Studios
Page no.	143	Mackintosh Studios
Page no.	144	Mackintosh Studios
Page no.	146	Joe O'Dell

"SAVE A LIFE, ENCOURAGE A FRIEND
TO BUY THIS BOOK."

Mail to:
PATHWAYS PRESS, 525 Jeffree Street, El Cajon, CA 92020

Please send me______copies of *"TRACKING: A BLUEPRINT FOR LEARNING HOW."* Single copies $12.95 (Calif. residents please add .78¢ sales tax) plus $1.00 mailing costs.

NAME____

ADDRESS____

CITY____________________________ STATE__________ ZIP_________

"SAVE A LIFE, ENCOURAGE A FRIEND
TO BUY THIS BOOK."

Mail to:
PATHWAYS PRESS, 525 Jeffree Street, El Cajon, CA 92020

Please send me______copies of *"TRACKING: A BLUEPRINT FOR LEARNING HOW."* Single copies $12.95 (Calif. residents please add .78¢ sales tax) plus $1.00 mailing costs.

NAME____

ADDRESS____

CITY____________________________ STATE__________ ZIP_________

"SAVE A LIFE, ENCOURAGE A FRIEND
TO BUY THIS BOOK."

Mail to:
PATHWAYS PRESS, 525 Jeffree Street, El Cajon, CA 92020

Please send me______copies of *"TRACKING: A BLUEPRINT FOR LEARNING HOW."* Single copies $12.95 (Calif. residents please add .78¢ sales tax) plus $1.00 mailing costs.

NAME____

ADDRESS____

CITY____________________________ STATE__________ ZIP_________

"SAVE A LIFE, ENCOURAGE A FRIEND TO BUY THIS BOOK."

Mail to:
PATHWAYS PRESS, 525 Jeffree Street, El Cajon, CA 92020

Please send me______copies of *"TRACKING: A BLUEPRINT FOR LEARNING HOW."* Single copies $12.95 (Calif. residents please add .78¢ sales tax) plus $1.00 mailing costs.

NAME ______________________________

ADDRESS ______________________________

CITY ____________________ STATE__________ ZIP__________

"SAVE A LIFE, ENCOURAGE A FRIEND TO BUY THIS BOOK."

Mail to:
PATHWAYS PRESS, 525 Jeffree Street, El Cajon, CA 92020

Please send me______copies of *"TRACKING: A BLUEPRINT FOR LEARNING HOW."* Single copies $12.95 (Calif. residents please add .78¢ sales tax) plus $1.00 mailing costs.

NAME ______________________________

ADDRESS ______________________________

CITY ____________________ STATE__________ ZIP__________

"SAVE A LIFE, ENCOURAGE A FRIEND TO BUY THIS BOOK."

Mail to:
PATHWAYS PRESS, 525 Jeffree Street, El Cajon, CA 92020

Please send me______copies of *"TRACKING: A BLUEPRINT FOR LEARNING HOW."* Single copies $12.95 (Calif. residents please add .78¢ sales tax) plus $1.00 mailing costs.

NAME ______________________________

ADDRESS ______________________________

CITY ____________________ STATE__________ ZIP__________